1906 – 2006

CENTENNIAL
Steaming Through the American Century

PHOTOGRAPHS AND NARRATIVE BY CHRISTOPHER WINTERS

Edited by Jessica Lothman

Bob and Sandy – Fair winds! Hope you enjoy the book...

3L-2s

— Chris

RUNNING LIGHT PRESS
www.runninglightpress.com

A Running Light Press Book

Photographs and Text © 2008 Christopher Winters

All rights reserved. Except for the quotation of short passages for review purposes, no part of this publication may be reproduced in any form without prior permission of the author.

1st Edition

1st printing May 2008

Published by
Running Light Press
1425 N. 65th Street
Wauwatosa, WI 53213
(414) 257-4168
www.runninglightpress.com

ISBN: 978-0-9814766-0-5 LCCN: 2008902690

Designed by Chris Winters and Jessica Lothman

Printed and bound in Canada
by Friesens of Altona, Manitoba

Pre-press design consultation and color management by Greg Dunn and Karen McDiarmid of Digital Imagery
www.digimagery.com

Individual prints of the photographs in this book are available from www.runninglightpress.com or by calling Winter Studio at (414) 257-4168.

Dust jacket cover: The S.S. *St. Marys Challenger* arriving at South Chicago, December 2005

Front dust jacket flap: Postcard view of the *William P. Snyder* departing the Canadian Sault, circa 1910

Front endpapers: Illustration of an ore boat under a battery of Hulett Unloaders from *This is Shenango*, 1954

Photograph of the *William P. Snyder* on the floating drydock at the Great Lakes Engineering Works, winter 1906

Back endpapers: Postcard view of the *William P. Snyder* unloading at Cleveland, Ohio, circa 1920

The *St. Marys Challenger* vanishes into lake steam, departing Milwaukee Harbor, December 2006

Dust jacket back cover: Afterglow on Lake Michigan, downbound off Sheboygan, Wisconsin

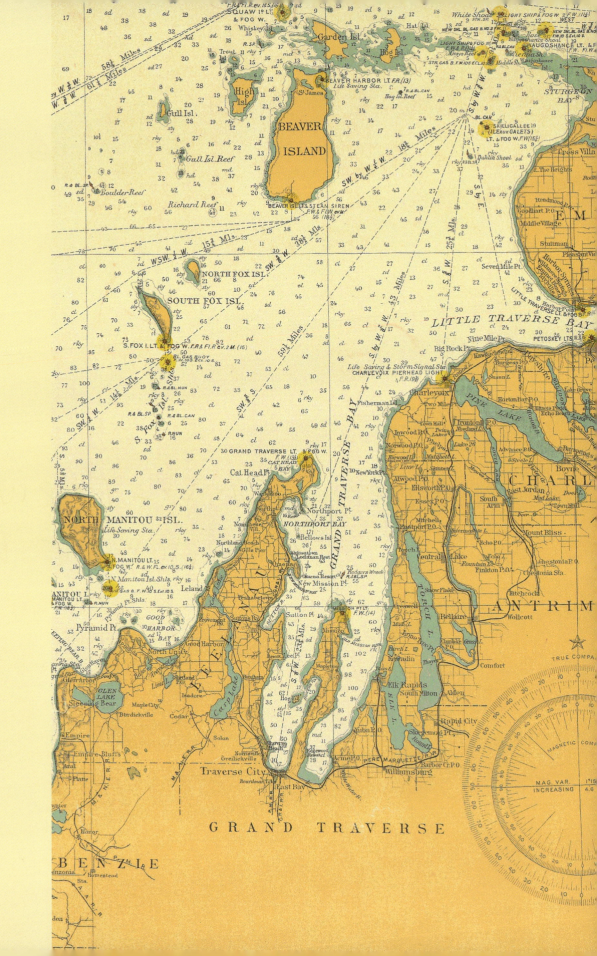

For Ed Hogan and my shipmates aboard the *Challenger*—real steamboat men.

S.S. COL. JAMES M. SCHOONMAKER   S.S. WILLIAM P. SNYDER JR.   S.S. SHENANGO   S.S. WILLIAM P. SNYDER   S.S. WILPEN

The Shenango Furnace Company fleet rafted together in Cleveland Harbor, April 23, 1917

Acknowledgments

First and foremost I'd like to thank my colleagues at Discovery World at Pier Wisconsin for putting up with the frequent, often open-ended absences enforced on me by the unpredictable nature of steamboats and photography. Humble thanks to our chairman Mike Cudahy—a kindred camera geek—who has kept me gainfully employed during the years it took to create this book. Ed Hogan, Aaron Bensinger, and traffic manager Bob Clark each deserve a knighthood for graciously putting up with my excruciating e-mail and cell phone pestering for five long years. Without their shared enthusiasm for this project, the *Challenger*'s remarkable anniversary would have gone unrecorded. To everyone in the Hannah Marine office now and then—I doff my cap. Special thanks to Norm O'Neill and everyone at the St. Marys Cement Group for granting me unfettered access to their magnificent machine.

Now that the *Centennial* effort has wrapped, I sorely miss speaking to someone in the Groh clan daily—at some junctures three or four times daily. Rocky, Barb, Peter, and Brendan… may God have mercy on my soul for the strain I put upon your collective cell phone bills. More than any other factor in this story, it was your willingness to cheerfully make and take phone calls at any hour of the day or night that made the photographs in this book possible. Thanks to Captain George Herdina and Chief Engineer Mike Laturi for making me feel completely welcome in their respective domains, and special thanks to Steward Sam Al-Samawi for packing me with sirloin and omelettes to the point of coma. Heartfelt thanks to all the officers and crew of the boat. I hope this effort stands as a tribute to your professionalism and genuine human decency; thank you all for welcoming a stranger into your midst.

Thanks to Art Chavez for his relentless cheerleading and beer bringing-over. Thanks to Roger LeLievre for letting me use his home office in Ann Arbor as my lower lakes base camp, and rallying me from the brink of despair on a number of occasions. Thanks to Bill Moss for sharing his terrific pencil sketches that grace this book throughout. Thanks to curators Matthew Daley at the University of Detroit–Mercy's Dowling Collection and Bob Graham at the Historical Collections of the Great Lakes for sleuthing out archival material and photographs. Seven thumbs up to Jim Bartke for sharing his Kodachromes of the Chicago River "jinx ship," and to John Belliveau who created a series of his meticulous color profile illustrations of the boat as she evolved through the century. Thanks also to Paul LaMarre III, who came to the rescue several times with arcane goodies pertaining to the Shenango fleet. Let me state for the record that I know just enough about writing and graphic design to be dangerous. Thanks to Molly Quirk, Lori Jacobsen-Tews, Greg Dunn and Karen McDiarmid for sharing their talents so lavishly and preventing my megalomaniacal vision for this book from rampaging out of control. Unlimited love to my folks, who have always stood unshakably behind my stroked-out projects. My grandmother did not live to see this book published—her interest in my travels around the Lakes always delighted me—and I know she would be pleased. The inestimable Capri swooped onto the scene in the nick of the moment, rekindling my own enthusiasm for this work, and delivering a strategic kick in the shorts when I needed it most. I'm pretty sure I've managed to leave somebody out… if you find yourself in a state of mild pique as a result, be sure to give me an earful about it, and as God is my witness I'll get busy baking you a big, chocolate apology cake.

Introduction

Asked to write a forward to this amazing book, I found myself humbled, not by the magnitude of the task, but by the history between these covers.

Rare is the volume that takes such an in-depth look at a single subject, especially when that subject commands an admittedly small audience. To say no one will be getting rich on the royalties from this labor of love is the absolute truth and makes all the more remarkable the author's passion for his topic. What a remarkable boat and what a remarkable book.

Even rarer, however, is the subject vessel of "Centennial," the honest-to-gosh steamboat *St. Marys Challenger*. Since just after the turn of the last century, predating the *Titanic* and before World War I, she's been earning her keep on the Great Lakes, a workhorse first hauling iron ore and coal, and now, Portland cement. At this writing, she's observing 101 years as a continuously operating steamship, the most senior of citizens in a category of vessels that have already disappeared on the oceans of the world and are dwindling in numbers on the inland seas.

Has any vessel anywhere in the world reached such a remarkable milestone? Certainly not on the Great Lakes. Other, older hulls survive, of course, but they no longer pulse with life, relegated to sidelines as storage barges. Not the *St. Marys Challenger*. When spring arrives, her boilers are lit once again, a head of steam is raised and finally, her helm controlled by the touch of a skilled wheelsman, the *Challenger* heads out to begin another season.

Under these remarkable circumstances, it would only be appropriate to christen this volume with a three-word invocation: "Long Live the Queen!"

– Roger LeLievre, August 5, 2007, on Lake Superior

Steaming Through the American Century

Waking up one morning and finding yourself obsessed with a floating dump truck is a strange universe to inhabit, I can tell you. It's a bit like having six toes–most of the time you'd prefer to keep your shoes on. The slide from polite enthusiast to full-blown mystic was a gradual one, and the fever manifested itself in small ways. For instance, at some point during the five years it took to collect the photographs for this book, it no longer seemed odd to doze off at a bridge gate, parked alone in a strange city at some godforsaken hour. Bridge bells are a sad and lovely thing to listen to in the bitter watches of the morning, their chimes threading through fitful patches of sleep, underscoring a scene that could have played out in 1915 or 1951 or 2006.

The Silent Stars Go By

From the 95th Street Bridge on Chicago's South Side you can see the fuzzy outlines of Upton Sinclair's raw Gilded Age. The vast appetites of this complex of industry on Lake Michigan's south shore clawed open the great iron ranges of Minnesota during the latter half of the nineteenth century, and gave birth to a lake-borne transportation system capable of filling up a continent. It's a used-up place, full of memory and regret and too much civilization. Lake boats still call on the Calumet River at South Chicago, the hard fragment that have survived the passage into the Information Age. As a point of information, somewhere it was decreed that anything that floats on America's five Great Lakes, from a hollow log to a 1000-foot super-carrier is tabbed a "boat." These giant boats arrive with stone and cement, and depart with coal for power plants on the Upper Lakes. Hardly sexy stuff, unless you number yourself among the odd fraternity of freshwater "boatnerds," as I do. So it was that I found myself in an unusual spot on Christmas Eve 2004, jerked awake by the cry of a steamboat whistle off in the darkness. Scanning around for any pipe-wielding drug fiends that might have been lurking in the shadows, I jumped out of the Honda and went to work–just camera shutters and bridge chimes and an exquisite loneliness.

The ice-rimed S.S. *St. Marys Challenger* appeared from a blue wall of lake steam hovering beyond the E.J.&E. railroad bridge. She coasted toward the still-shuttered 92nd Street bridge, whistling for the tender–who was evidently asleep or preoccupied with some last-minute gift wrapping. Inbound for storage silos about six nautical miles upriver from Lake Michigan, she'd arrived here on the last leg of a routine voyage from Charlevoix, Michigan, with Type I Portland cement. For a true believer like myself, behexed at the age of nine by the mythology surrounding the wreck of the *Edmund Fitzgerald*, the sight of this phantom crinkling through the skim ice was sublime–and uniquely my own, as nobody else in the metropolis was mad enough to be huddled on the banks of the Calumet on Christmas Eve to witness it. Wreathed in billows of steam exaggerated by the arctic cold, she slipped through 95th Street, then the "Five Bridges," section of the river, letting off another whistle blast over the ruined landscape. Bridge bells started up at 100th Street. I bagged the camera and cut over to the south, tucking the car on a side street and jogging up to the river bank with my gear in tow. Three westbound cars were penned up on the bridge approach, a common syncopation in the rhythm of life in South Chicago. I marched up and back on the sidewalk beneath the tender's tower, firing the shutter as best I could with cold-scalded hands. The driver in the lead car stared at me like I had a penguin on my head. I couldn't help wondering if a single one of the bleary travelers marooned at the bridge gate apprehended the fact that the machine creeping through the draw was a rare and spectacular monument to everything great about the American Midwest. Trying to explain *Challenger*'s mystique to a civilian is a bit like cutting roast beef with a screwdriver.

I slammed back into the car, throttling the heater full on, and zipped south on Avenue N, ahead of the boat. The bascule bridge at 106th Street–sixth of nine spans *Challenger* would pass through en route to her destination upriver–squealed a bit in the cold, tilting up at the frozen stars to let the boat slide beneath. Bridge bells carried over the icebound river, floating off to dwell with the chimney vapors and jet contrails. In an age when the obsolescence curve of most products (including my digital cameras) is measured in months, it was comforting to follow this relic of America's industrial past from port to port–strange hours and ceaseless travel, all of it somehow familiar. A steamboat life is lived in moments like this one–a solitary Christmas Eve on a deserted river, and my life was lived in such moments for five years. I felt relevant perched on the 106th Street Bridge, frostbitten shutter finger scissoring up the scrambled light as we went about our necessary work, the boat and I. It was Christmas Eve, America was still the warm center of the universe tonight, and all was right with the world. Wrapped in her gorgeous plumage of white steam, with a humble string of Christmas lights winking in the officer's dining room, even the bloody Grinch would have had to admit that the ancient tub centered in my viewfinder had *heart*.

The Long Ships Passing

Descended from Captain Eli Peck's prototypical steamer *R.J. Hackett* of 1869, the "laker" was once a unique architecture in the world of ships. Inspired by necessity, like the McCormick Reaper and the Model T, the boats were an ingenious product of the heartland, and during the first Golden Age of the lake fleet between 1900 and the Great Depression, emblematic of the hog-wallow capitalist zeitgeist of the time. Like the Reaper and Model T, both forged from the same raw materials lake boats hauled, the evolution of these colossal vessels helped transform the texture of modern life in the first half of the twentieth century. On a "standard 600-footer" of the day–the class to which the *St. Marys Challenger* could loosely be assigned when launched– form followed function throughout. Configured with a high bow and pilothouse forward, engines aft, and a huge, uncluttered main deck, lakers of that era were engineered to patiently cover distance, sipping fuel at a modest speed, then squat under newly-developed unloading rigs where their cargoes were dug out with unparalleled efficiency. In an age before transverse bow and stern thrusters, these distinctive "double enders" evolved to navigate more easily through the Great Lakes' complex interconnecting web of dredged entry channels, locks and river systems.

Here at the dawn of the 21st century, enthusiasts define classic breeding in a lake boat as one of this traditional double-ended configuration. Advances in engineering and propulsion have rendered their unique profile obsolete–those built after 1971 aren't dramatically different in appearance than bulk carriers anywhere else in the world. For shore dwellers apt to pay attention to such things, this dwindling band of traditional lakers have become like giant citizens of the region, steaming past docks and boathouses once or twice a week for fifty, or seventy-five or a hundred years. They are *our* ships– built here in the middle of a continent–living, breathing extensions of our native genius. Our *boats*, excuse me; even the noun is squarely no-nonsense and reliable. The lake fleet bore a century of the Midwest's commerce on its broad back, helped win two world wars, muscled Communism into the ditch, and delivered the basic building blocks of civilization to a republic in transformation from rural outpost to global superpower. The flood of red ore and other bulk commodities that began in the 1850s has ebbed away, but you can still hear the echoes of a fleet that once numbered over 500 carriers and moved hundreds of millions of tons of cargo in a single ice-shortened season—more than all of America's saltwater fleets combined. That's a mouthful, sure enough. And here's another: Travel anywhere else in the world on Christmas Eve, and your odds of witnessing another 98 year-old steamboat delivering a paying cargo under her own power are about the same as keeping a lunch date with Stonewall Jackson and the Abominable Snowman.

Lure of the Lakes

By the gauge of any respectable steamboat enthusiast I was born too late. "Enthusiast" is, of course, too polite of a word to describe the fever. An enthusiast doesn't neglect food, sleep, friends and personal hygiene the way I did for five years at the dawn of the new millennium. I started shining around docks with camera in hand in the late 1990s, after much of a once-epic subject had withered away to shadows. By then, my homeport of Milwaukee, Wisconsin, on the west shore of Lake Michigan, was visited by just a handful of vessels a week. Gone were the year-round railroad carferry operations, saltwater tramps, and ancient whaleback tank ships calling at the municipal piers. Gone were the coal-fired, triple-expansion hand bombers of the Huron Cement fleet that crept to the head of navigation on the Burnham Canal. Gone were the Menomonee River haunts of the old-school self-unloaders *Nicolet*, *Hennepin*, and *W.W. Holloway*. Stone docks and bridge cranes weathered away in the Menomonee Valley as the river silted up and became unnavigable—a river that once fed the beating heart of a crucible regarded in its prime as Workshop to the World.

W.P. Snyder

There was atrophy and chaos in harbors all around the Great Lakes in the late 1980s and early 1990s, a sea change that infused every frame of film with a sense of historical imperative. Looked at from this end of the kaleidoscope, the real spark of my mania was David Plowden's landmark effort *End of an Era*, published in 1992, which subconsciously primed me to go for broke should a worthy quest ever present itself on the waterfront. Virtually every boat between the covers of Plowden's elegy to the last of the coal-fired lake steamers was cut to bits or rendered up to duty as a dead-eyed storage hull by the time I got my hands on the book in 1999. The majestic gravity emanating from these vanishing giants gradually peeled me out of the responsible orbit of a half-baked commercial photographer and into an eccentric universe of mingled art and history, a world of solitude and inestimable joy. A career—if you can call it that—with no sane commercial incentive and no real strategy behind it, but one steeped in irretrievable one-shot magic. I didn't realize it at the time, but when the *Southdown Challenger* (as the boat was known from 1999-2005) showed up at the Bay Shipbuilding Company one sunny evening in 2003, I was poised to take my best shot at a once-in-a-lifetime opportunity.

Ghost in the Machine

"Until one is committed, there is hesitancy, the chance to draw back, always ineffectiveness. Concerning all acts of initiative (and creation), there is one elementary truth the ignorance of which kills countless ideas and splendid plans: that the moment one definitely commits oneself, then providence moves too. A whole stream of events issues from the decision, raising in one's favor all manner of unforeseen incidents, meetings and material assistance, which no man could have dreamt would have come his way."

–William Hutchinson Murray
From *The Scottish Himalayan Expedition*, 1951

The book you are holding was never something I officially set out to create, which is probably a blessing. If I had known what was coming, the moment of its inception would have seemed a bit like shoving off from shore with the intent to row across the Pacific Ocean. In July 2003 I trundled my view camera up the gantry crane at Bay Shipbuilding in Sturgeon Bay, Wisconsin to photograph the World War II-era Coast Guard icebreaker *Mackinaw*. Big *Mack*–once the most powerful and sophisticated ice crusher in the world–was slated to be decommissioned as soon as her replacement had been delivered from the shipyard. This would be the last chance to expose a few sheets of film of her unique, below-the-waterline characteristics–an opportunity which rated putting up with the six-hour round trip to Sturgeon Bay and back.

After an hour or so aloft, I started to tear the camera down and mark up my film holders, when who should appear on the horizon but the venerable *Southdown Challenger*, headed for her own appointment with Bay Ship's drydock and the final phase of a decade-long bottom rehab. I got off a handful of shots of the 97-year-old boat arriving at the yard's fitting-out berth under tow of the tug *Bonnie Selvick* before I clambered down from the crane. I fired off a frame or two dockside, and on my way out of the yard–almost as an afterthought–I arranged to pay a return visit to the drydock later in the week, once *Challenger* was high and dry on the hull blocks.

It occurred to me in the three days that elapsed between visits to Sturgeon Bay that there might be something magic afoot. The boat had been idled for its Coast Guard mandated five-year hull survey, and considering the amount of money being invested in completing the bottom work at that same time, it was safe to assume her owners had every confidence in her re-certification. Extrapolating from there, and barring the totally unforeseen, it was not ridiculous to hope the boat might eclipse the magic 100-year milestone in 2006. That milestone being shattered and sinking to the lakebed like an anvil was an occasion Great Lakes boat geeks (myself included) had been waiting for like something prophesied in the Old Testament. The ancient *E.M. Ford* had come close, resolutely steaming over the lower lakes, pushed by her original quad-expansion steam engine, until 1996. Close but no cigar. As it turned out, she was banished to living death as a storage hull in Saginaw, Michigan, at the end of her 98th season.

The first four images I made of the boat that summer were recorded in traditional style using large-format transparency film. The rig I hauled up on the gantry crane and into the baking-hot engine room of the *Southdown Challenger* three days later consisted of a 4x5 inch monorail view camera, and five prime lenses. My hardware in July 2003 was fundamentally the same clumsy, tripod-bound boat anchor as Matthew Brady might have used on the Gettysburg battlefield in July 1864. Film stocks had evolved mightily over the decades, and modern optics were superior in their ability to render color and contrast, but nearly a century and a half after Brady's day the garden variety view camera still consisted of a simple front and rear standard, one to support the lens board, one to support the film plane, and a ground glass viewfinder. These vertical frames were geared to a rail that could be mounted on a tripod or other camera support, and the space between the lens and film plane rendered light-tight by a vinyl bellows.

Working with a view camera is a clumsy affair, framing the scene on the low-contrast ground glass backwards and upside down, then focusing the image with a 4x magnifying loupe, all the while tucked under a "horse blanket" dark cloth. Exposure calculations are dialed in with a hand-held light meter, and film is exposed one sheet at a time–often after some groping interpolation based on murky Polaroid proofs. When the light is changing or when shooting a moving subject the old school large-format photographer resembles a tormented mechanic ducking in and out from under the hood of a burning car, wild-eyed, muttering profanities and feverishly ticking off the seven or eight steps necessary to expose a sheet of cut film properly. Shooting in that mode–especially outside of a studio setting–was almost always a bit of a public spectacle, about as far removed from inhabiting a scene unobtrusively as you could get without actually wearing puffy director's pants and a beret.

The paradigm shifted dramatically in autumn 2003. That October, after Ed Hogan, V.P. of Operations at HMC Ship Management, gave me the green light to expand on the work I had created in the shipyard, I ponied up a staggering amount of money–more than I had paid for my beloved Civic hatchback, in fact– to get my hands on Canon's new digital flagship, the EOS 1Ds Mark I. The EOS Mk I (and its 16.7 megapixel successor, the 1Ds Mk II) were to me what the availability of affordable color film after World War II was to photographers of a previous generation. Boasting a quantum-leap, superbly low-noise full-frame sensor, the 1Ds Mk I was the first camera that convinced diehards everywhere to walk away from film forever. Canon's breakthrough CMOS chip technology was amplified by a new class of "image-stabilizing" zoom lenses, revolutionary large format ink jet printing technology, and the ever-evolving miracle of Adobe Photoshop software. Photography was in the throes of a run-like-hell-for-the-top-of-the-next-hill era, and became a very exciting calling to be wrapped up in at the dawn of the new millennium.

Now, I had not laid a finger on a computer until my 30th birthday in 2001. But a wild sense of enthusiasm for these powerful new tools aided in pushing through the technophobia… and took the sting off painfully short obsolescence curves early on. *Painfully* short. After Canon dropped their 12 megapixel "prosumer" EOS 5D camera in 2005, I was only able to recoup $1,500 for the 11 megapixel 1Ds Mk I which had cost $8,000 brand new less than two years prior. Adding insult to injury, my Hasselblad film equipment–formerly the gold standard–was so devalued in this new digital universe that the internet broker whose services I used to buy and sell equipment wasn't even interested in trying to peddle the Hass outfit for me… simply not worth his time. In one sense it was a gruesome scene, especially if you were working in a niche vein that promised little in the way of return-on-investment. BUT (and this is a stupendous but) you could now shoot with profligate National Geographic staffer abandon all day, archive and edit practically in real time, then dump the flotsam and jetsam into the Twilight Zone with a single mouse click. If you liked to really *cover* your subject, as I did, the digital camera paid for itself in less than 18 months based on film and processing savings alone. No more basements packed with moldering slide pages, and no more putzy film scanning. Manufacturers were quick to produce an affordable and reliable memory card the size of a fat postage stamp that boasted as much data storage as my first desktop computer. Factor a laptop computer into the equation, and the instant gratification–unfettered by anything but the occasional need to access an A/C power supply–was utterly spine tingling.

Steamboat Fever

I made my first trip aboard the *Challenger* in November 2003. For the mingled soul of a boat geek photographer, it was a five-day thrill ride at 10 knots. Pushed to the limit, the 1Ds Mk I's sensitivity to light was five times greater than the transparency film I had been shooting with to that point. With a little Zen concentration, and a jury-rigged shoulder stock of sorts, I was suddenly able to squeak out images on a moving platform in the virtual dead of night. I was able to capture full color images of engineers and oilers at work in the nether regions of the boat, where mixed and muddy lighting sources would have driven my daylight-balanced slide film right around the bend. A powerful but tiny battery-powered strobe punched in a little clean light where needed. The camera's color LCD screen allowed me to proof my composition and exposure much more fluently than a Polaroid ever did, without generating the noxious chemical waste and crushing expense. If I clicked on the tack-sharp 70-200 mm gyro stabilized telephoto lens and padded around, I could observe without being observed. It was a brilliant thing to be able to work so spontaneously in available light after years of being tethered to an entire hatchback full of equipment.

On hearing me spell out the contrast between old and new technique, a friend commented that it sounded to her like dancing a whole tango in one step. The blasphemous thought occurred to me on the road home from Charlevoix after that first time out, that here was the logical extension of Plowden's elegant old-school B&W masterpieces in *End of an Era*. Now, that was some breathtaking smack talk: to presume to build where a great tradition already exists, to presume to record the last echoes of an old way of life in a brand new way—and with a little luck—be present at the smashing of the numinous 100-year barrier. Punter or not, if I failed to pull it off it wouldn't be for lack of trying. From the minute we cleared the breakwall at Charlevoix on that first voyage (after four pitiless, sleepless days chasing the boat all over the gale-swept lower Lakes), and for five years thereafter, I approached the project as the work I was born to do.

By the time *Challenger* fit-out in April 2004 I had accumulated just enough professional experience and digital equipment to be dangerous. My warhorse Honda had 150,000 miles on it, but with some fresh tires and a new transmission it would see me through. A freelance arrangement with Milwaukee's evolving Discovery World museum provided a steady income stream, and just enough flexibility to pursue the mad puzzle of tracking and shooting a moving target across three of the five Great Lakes—and to endeavor to do so under truly picturesque weather and lighting conditions. Real-live steamboat geeks in the Hannah Marine office continued to bless the effort with the rarest of carte blanche access to the boat for as long as it took for me to call it finished… even after the paper monster of post-9/11 MARSEC Homeland Security legislation got cranked up in July 2004.

Most importantly, to my great relief and occasional delight, after a trip or two the boat's 24 person crew seemed to get on board with the fact that a strange dude was going show up and begin blasting away over, and over, and over again— at least nobody ever hit me upside the head with a wrench and tossed the camera overboard. I'm sure I was spared this fate because *Challenger*'s first mate took me under his wing and made sure I didn't commit too many subcultural faux pas. Ray "Rocky" Groh— a fellow artist and a fifth generation lake sailor, became a dear friend over the course of the project; with cell phones at the ready the entire Groh clan—Rocky, Barb, Peter, and Brendan—became the primary conduit in my cat-herding apparatus.

When navigation opened in the spring of 2005, I was ready with the 16.7 megapixel Mark II version of the 1Ds, and was soon editing images that were as tack sharp and nuanced as earlier medium-format work— while underway in the middle of Lake Michigan. The incredible audacity! I built half the page layouts for this book out of the sight of land. For someone who had been schooled in traditional offset printing and chemical darkroom techniques, E-6 lab bills that read like the tab for an aircraft carrier, and $150 C-prints, this new era was like being photon-beamed into a science fiction movie. In the span of less than five years, if you were a sincere fool with a niche subject and a willingness to learn how to wear several hats, you no longer had to sit at the back of the bus production quality-wise. It was a revolutionary time, and the irony of using tools forged in the last years of the 20th century to meditate on a revolutionary machine created in the earliest part of it was not lost on me.

A Giant's Tale

In November 1905, when the *Challenger*'s keel plates were laid down at the Great Lakes Engineering Works' Ecorse, Michigan shipyard, Teddy Roosevelt was in the White House, milk cost seven cents a gallon, and the R.M.S. *Titanic* was still a shipbuilder's fantasy. On February 17, 1906, when Hull No. 17 dropped into the icy Rouge River, men wore spats to the launching party and talked of the profound impact the horseless carriage was having on Detroit and the country at large. As long as a city block, the boat was one in a new class of steel giants muscling older, wooden-hulled carriers out of the lucrative iron ore trade. Christened *William P. Snyder* by Ms. Mary B. Dyer of Pittsburgh, the boat was the first of five ships built for the Shenango Furnace Company at the turn of the 20th century. Though never quite a record-breaker or a beauty queen, it is fitting somehow that No.17 has survived to become the last living monument to a monumental class of ships, as her blood is bluer than most.

Named in honor of Shenango's founder and president, the *Snyder*, weighed in at a whopping 552 feet in length overall and 56 feet of beam, with a molded draft of 22 feet. The boat's namesake, Mr. William P. Snyder, was a blue-blooded descendant of Pennsylvania's first governor, and one of the buccaneering capitalists who opened Minnesota's Mesabi Range–the mother lode of all iron ranges–to exploration in the late nineteenth century. One of the largest vessels of her day, and an engineering showcase, the boat was outfitted with oak-paneled guest quarters and an observation lounge that were the last word in Victorian exuberance. In an era before anyone could pack an overnight bag and jet off to the Caymans for some R&R (and pre-dating air conditioning), it was considered quite the dog day tonic for a VIP client or favored employee to steam off on a cruise to the head-of-lakes and back. Driven by a 1600 horsepower, triple-expansion reciprocating steam engine and two Scotch boilers, she was 10,300 gross tons of burden in three cargo holds, served on her main or "spar" deck by 31 hatches on 12-foot centers. Shipping concerns on the Lakes had been leapfrogging each other in Wild West style for a decade and after a brief flirtation with greatness, the *Snyder*'s handsome lines were swiftly eclipsed by finer and more capacious lakers–including her fleetmates *Wilpen*, (which boasted the Great Lakes' first and only steam-powered orchestrian piano in her observation lounge) and big sister *Colonel James M. Schoonmaker*, which at 617 feet in length and 15,000 tons capacity was the largest bulk carrier in the world when she came out in 1911. The *Schoonmaker* remains the only lake boat before or since to have laid claim to a "world's greatest" title. Idled for good in 1980, she too survives to this day, enjoying a stately retirement on the Toledo, Ohio, riverfront as the museum ship *Willis B. Boyer*.

Tricked out in green and white Shenango livery, with her signature orange boot-top visible at the waterline, No. 17, known officially now as the S.S. *William P. Snyder*–bowler-topped captain Henry Peterson commanding–departed Detroit on her maiden voyage to the lakehead for a cargo of red ore on April 28, 1906. More than a century later, while laying for weather off St. Ignace, Michigan, in October 2006, the boat's relief mate Bill Kischel and I were shooting the breeze over a mug of steamboat coffee. Kischel seemed to resonate with my overripe historical predilections more than the rest of the crew–many of which genially wrote me off as a well-meaning kook.

We began waxing poetic about the durability of the steadfast tub under our feet, how many faces and presidents and wars, and how much pounding and twisting and evolutionary change this old boat had weathered, trying to fashion a prism to tease apart the color of her unique pedigree in the world of ships. At some point a light bulb went off over Kischel's head, and he jabbed a finger out of the chartroom window. Below us, one of *Challenger*'s three deckhands was coiling a heaving line on the steel deck–the youngest hand on the ship's articles, the mate observed, flipping through his personnel index cards. That small fact sparked a line of conversation that proved one of the most memorable in my experience. We determined that the youngest deckie aboard had been born 21 years after America's first space walk in 1965. Kischel finished coding the afternoon weather report, and began puzzling out another calculation. My eyes scanned the darkening water, where a quarter mile distant the deck lights of the old rock boat *Calumet* began to sparkle. Kischel tapped the notepad to reveal his simple arithmetic, and in its modest sum we found our prism. We caught a glimmer of the vast accretion of hours and faces, a century of world events and a million miles of water gone under the keel of this remarkable machine. The mate stepped out on the bridge wing to have a smoke, and I was left with my thoughts and the coffee, which was brewed strong enough to float an iron wedge. The unassuming heaving line that sparked our whole epic bull session lay on the same steel deck trod by captain Henry Peterson of Detroit–*Challenger*'s first skipper–who was born 20 years before the end of the Civil War in 1865. Clearly, we were afloat on the most extraordinary of ordinary ships.

Live Steam

During the palmy days of the lake fleet in the early 1950s, the casual watcher on the beach could have witnessed a long ship slide past Windmill Point on the Detroit River once every 60 seconds. In that era over 400 ships, big and small, routinely traded in and out of Lake ports on the American side. By the outset of what I began to loosely refer to in e-mail correspondence as the *Centennial* project in summer 2003, the U.S. registry of vessels in active service on the Great Lakes had shrunk to just 53 (granted some of them boasting six times the capacity of a standard 600-footer), and it would have required a vivid imagination indeed for our casual watcher to imagine that 48 ships had once called on Duluth harbor in a single day.

A profitable second career as a dry bulk cement carrier from 1967 onward allowed *Challenger* to weather the holocaust of the 1980s, when economic forces–both global and domestic–came home to roost, and hundreds of lake vessels were led to the bone yard– some 50 years younger and three times as capacious. It is a lucky thing for our purposes that Portland cement is a picky cargo, and that the boat was rebuilt to move it with unparalleled efficiency.

In the spring of 1950, Hull No. 17's remarkable career on freshwater was bolstered by the mechanical equivalent of an open heart surgery. At the tender age of 44, her original triple-expansion steam engine was replaced by a 4-cylinder Skinner Unaflow steam plant at the Christy Corporation yard in Sturgeon Bay, Wisconsin. The repower was notable in its day, as Christy developed and used for the first time a cofferdam system that allowed wet-side work to be done without the boat having to pay an expensive visit to the drydock. The Unaflow Marine Steam Engine, most built by the Skinner Engine Company of Erie, Pennsylvania, is generally regarded as the last unique genus of steam propulsion on the world's waters. They were widely utilized after World War II to repower lake boats, who served a trade where high speed wasn't necessarily an issue. That being said, the curtain had been coming down on the age of steam for decades–even on freshwater where long service lives tended to slow the evolutionary curve. The last boat repowered with a steam engine on the U.S. side of Great Lakes was the *Charles M. Schwab* in 1961. Inexorably, as the last bastion of teakettle technology began to crumble–as it had already on the world's oceans–the ear-splitting diesel engine would become the power unit of choice on lake boats.

Now, the average diesel engine has about as much romance associated with it as a lawn mower. Fortunately, No. 17's engine is an oily blued-steel thing of beauty, and one of the last of its kind. *Challenger*'s Erie-built Skinner is left-handed (which is to say the engine turns the shaft and propeller in a counterclockwise rotation) and it smells wonderful–faintly like a movie theater popcorn machine, I always thought. It's loud of course, but instead of a diesel's explosive drone, this engine thumps out a quick, primordial rhythm, as its four pistons intake and exhaust high-pressure gouts of steam. Backlit view ports in the engine cowling allow watchstanders to observe the circulatory pulse of lube cams and rollers as they whirl, and portholes on the crank deck permit a glimpse of the engine's business end, where bare light bulbs inside the engine frame throw a gloss on the brute steel of the crankshaft as it flies around flinging oil, driving the propeller and all 17,000-odd gross tons of the boat and her cargo forward. The plant is an engineering landmark of sorts, being one of only four such engines still functioning in the Western Hemisphere, and the only simple-type Erie Skinner still earning a paycheck. Two live Skinners reside in another denizen of Lake Michigan–the 55 year-old carferry S.S. *Badger*–whose twin compound steeple-type engines are designated historic landmarks by the American Society of Mechanical Engineers. No such plaudits for *Challenger's* beating heart as of yet, which thunders on in anonymity, tended to by a veteran nucleus of engineers under Chief Mike Laturi who grimace and sweat over the engine's every mood… knowing that a serious mechanical meltdown will presumably spell the end of steamboat life as they know it.

As a guy who can barely fix a leaky toilet, I always reserved a little hero worship for these engineers, who could build anything, fix anything, and string together the finest cuss words in eight states and two provinces. Battling 90° F. temps on any given watch, the guys performed man-sized work–beating their blackened cut-sleeved overalls to rags in the process. When they weren't baby sitting the antique engine and its scalding-hot two story boilers, idle moments filled up dealing with the cranky plumbing and steam radiators in every living space on the boat, maintaining what amounted to a high-volume restaurant kitchen in the galley, and grappling with the unloading system's complex train of rollers, buckets, blowers and dust collectors. Any hour of the day or night you might find them busting out the air tools if the boat's venerable steam winches started acting ca-ca, or if the bowthruster got a hair across it's you-know-what, and if–God forbid–something went wrong with the satellite television system, they had a real crisis on their hands.

First Assistant Engineer Dave Jarvis once tried to explain how the engineering department grapples with the care and feeding of the boat's existing vacuum-tube electrical plumbing and an aging solid state boiler automation system. I won't bore you with the details, as I understood about every tenth word of it. Suffice to say, it demands the occasional feat of Yankee ingenuity. Such feats are taken as a matter of course in what amounts to one of the most specialized and professionally challenging engine rooms on the high seas. Third Assistant Kevin Rogers tried and failed during three separate tutoring sessions, complete with idiot-proof diagrams, to explain how a Unaflow works its mechanical magic, with a complex maze of one-way valves and poppet springs. I glazed over every time, but he did manage to communicate the origins of *Challenger's*

most unique mechanical and aesthetic accoutrement–
her glorious and ever-billowing plumage of steam.

Challenger's steam-powered auxiliary systems are unique
on the Lakes. Low-pressure steam is bled off the engine
to power deck winches, generators and other assorted
cogs in the equipment train that drives the boat. This
live steam does its work, and is exhausted overboard
through a vent pipe in the smokestack. Not the most
efficient use of energy perhaps, but very, very lovely to
look at. As long as the steam plant is on line, this plume
hisses and writhes over the after end of the boat, and
provides a fabulous photographic companion. It *lives*. A
jet of pure white steam, the plume shifts shape and size
depending on wind and temperature… it howls, snorts
and subtly changes color depending on the hour of day
and the prevailing weather. It really goes ballistic when
the temperature drops–and the jet expands violently
as it hits the frigid air. I quickly learned a steamboat is
never more alive than in sub-freezing weather, as the
beast billows, howls and sheathes herself up in frost
feathers. A seasoned eye can recognize the ancient
tub from 15 miles distant on account of this utterly
fantastical rooster's tail.

Any sailor will tell you that a ship is an anthropo-
morphic creation to begin with, a living machine
that has a name and a face, and a personality–some
to the good, some to the bad. A laker smokes and
smells, whistles, cries out, flexes and moves… for me,
Challenger's magnificent steam plumage was capable
of propelling that sense of living tissue, of steel skin
stretched over a beating heart, into overstimulated ship
geek hyperbole. The steam pouring out of her stack
was like an arrangement of notes that would never be
played again. It was hypnotic and unforgettable, and
every once in a while I would get caught by one of the
after-end guys just staring at it, glossy-eyed, as if I kept a
big stash of peyote in the owner's lounge. That could be
a bit mortifying.

The Turning of the Tide

By the early 1960s, the iron ore trade on the Lakes was
being serviced by a new class of post-war giants in the
25,000-ton range, and more and more coal was being
moved in a fleet of hulls converted to the high-efficiency
self-unloader configuration. Prospects began to look
rather grim for welterweight No. 17–60 years old in
1966, an increasingly undersized senior citizen, even
by freshwater standards. To sum up her lineage to
this tipping point: the boat hauled for the Shenango
Furnace Company as the *William P. Snyder* from her
maiden voyage in 1906 until peddled to a subsidiary
of the Pickands Mather fleet in 1926. She sported an
umber hull and orange stripe herald on her stack as
the *Elton Hoyt II* for Pickands until 1951. During this
incarnation her triple-deck pilothouse was cut down in
favor of a more modern forward-end configuration, and
her paneled guest quarters were stripped out to make
room for additional crewmen when the industry went to
a three-watch system after World War II. In 1951, after

re-powering, she swapped identities within the Pickands fleet, becoming the *Alex D. Chisholm*, and continued to haul ore and coal for what became the modern incorporation of the Interlake Steamship Company.

It looked like curtains for No. 17 by 1966. Then, as the fickle steamboat gods would have it, she met a swift reversal of fortune, and was re-engineered for a second career which as of this writing has spanned some 40 years. Salvation emerged in the unlikely form of the Medusa Cement Company of Cleveland, Ohio. Concrete is so rarely associated with things that float. However, the conversion of the laker *Samuel Mitchell* to a self-unloading cement vessel in 1916 pioneered the economical movement of bulk cement by water. The success of the *Mitchell* lead to the commissioning of two lake vessels designed and built specifically to haul cement, the *John W. Boardman* of 1923 (better known as the *Lewis G. Harriman*) and the *S.T. Crapo* of 1927 (a most unfortunate name for a boat, I know).

Within a decade cement was officially added to the list of the bulk commodities that were floated in great quantities on the Great Lakes. Cement–literally the glue that holds together the sand and aggregate in concrete– was used extensively by the Greeks and Romans, held the Great Pyramids together, then was lost to civilization during the Dark Ages. Natural cement slowly made its way back onto the world stage during the sixteenth century. In 1824 British stone mason Joseph Aspdin patented his formula for what he tabbed "Portland cement" and laid the foundation for an industry responsible for churning out the most widely used building material in the world today–concrete.

Like any of the bulk trades on the Lakes, the commodity in question is not sexy. In fact, talking about Portland cement at any length will lull a tax attorney to sleep– but just try building a modern world without it. Without a reliable means of making and delivering Portland cement, roads and bridges would still made out of logs.

No more pre-stressed concrete buildings and stadiums. Try building an interstate highway system out of adobe bricks sometime. It's unthinkable.

Purchased by Medusa in June 1966, the boat was towed to the Manitowoc Shipbuilding Company after being idled by Interlake at Erie, Pennsylvania. Stouthearted No. 17 was to be rebuilt in the specialized role of a cement carrier. Medusa was opening a new state-of-the-art quarry and kiln at Charlevoix, Michigan, in 1967, and needed to get product to its distribution centers around the Midwest. Her modest dimensions, formerly the seed of her gathering doom, worked to her advantage when viewed in the context of her new trade. Ore, coal and lumber docks had been established on the prime real estate of most Great Lakes harbors beginning in the middle 19th century. Cement was a latecomer to the bulk scene, and as result most storage silos were located in the cheap seats; in the case of many of Medusa's seven shoreside terminals, at the bitter end of navigation on several twisted tributary rivers. A smaller boat could put up better numbers in this niche trade. An uber-efficient cargo handling system was also deemed to be a critical asset, as Medusa would have only one ship to haul product to all seven silos. Handling a fussy cargo faster was the key attribute engineered into *Challenger* in 1967 that saved the day when it came to shattering the 100-year barrier four decades later.

So it was in 1966 that the boat was rebuilt with a landmark system of air slides, a centerline conveyor and bucket elevator that, mated with similarly muscular shore side equipment, allowed her to unload Portland cement faster than a jackrabbit on a date. The former *Alex D. Chisholm* was given newly automated boilers and re-christened *Medusa Challenger* during a riverfront gala in her new homeport of Cleveland, Ohio. The boat could discharge her finely-powdered cargo five times faster than any of her competitors–a rated 1500 tons per hour, versus an industry average at the time of 300 tons per hour. She had been magically transformed into a 60-year-old thoroughbred of sorts, and kept steaming profitably through the latter half of the 20th century.

From 1967-1980 she was the largest vessel transiting the Chicago River, creeping inbound and out through the glass canyon of The Loop, up the river's North Branch to Penn Dixie's silo on Goose Island. Befitting her mythological namesake, the *Medusa Challenger* became notorious for turning the city's legion of bridges to stone. In the thick of the her transit downtown, as many as three bridges would be open over the boat simultaneously, and all three might be stuck open if an oncoming bridge was frozen shut. This "jinx ship" phenomenon created such heartburn during rush hour traffic that the city lobbied to relocate the cement silo to a new site on Lake Calumet. My only regret during the *Centennial* project was that I was never able to record this extraordinary voyage into the heart of the City of the Century with the latest generation of digital camera equipment.

Beginning in 1991, Medusa saw fit to invest millions in a comprehensive re-skinning campaign, with shipyard gangs progressively re-plating the boat's lapped and riveted hull–then 82 years old–from light waterline to light waterline, one end of the keel to another. It was considered money well spent, as *Challenger* is put together like a battle tank with framing spaced on 39-inch centers. By comparison, a modern ship built to replace her would have frames spaced on 8 to 10 *foot* centers. The face lift was projected to extend the hull's useful life by another 30 to 50 years.

Business as Usual

After decades of status quo, things got a bit sketchy for *Challenger* as the curtain came down on the 20th century. The Medusa Cement Company was bought out by UK-based multinational Southdown Cement in 1999, during what proved to be a heaving era of consolidation in a globalized industry. The boat was renamed *Southdown Challenger* at fit-out that year. Southdown was gobbled by Mexico City-based giant Cemex in 2002. Cemex never bothered to change the boat's name, and sold its interest in Medusa's former Great Lakes assets to Brazilian giant Votorantim in early 2005. The boat fit-out unusually late that spring, and a new dynamic was introduced into the plot line–Votorantim's North American subsidiary St. Marys Cement already had a fleet of tug-barges to move its product around the lakes.

As soon as the Cemex buy-out was announced, dark speculation arose about the boat's continued health and well being, as a result of overcapacity and other merger-related ennui. Good 'ol No. 17 fit-out unusually late in 2005–in mid June–steaming into her 99th season under a sixth name, the *St. Marys Challenger* (no apostrophe on "Marys," a native appellation). The boat was absent on typical trade routes to Detroit, Toledo and Cleveland for the first time in decades, servicing four former Medusa terminals on Lake Michigan exclusively.

Would there be enough work to go around for what had suddenly become a five-boat stable? *Challenger*'s fleet-mates were all tug-barge combos with newer machinery and lower manning requirements–thus lower payrolls. Business is business after all. Numbers were crunched, blueprints drawn up, and conversion to a barge (the steamboat equivalent of a full-frontal lobotomy) at the end of the year began to look imminent. The 2005 shipping season played out like a slow-motion drama, with new rumors about the boat's fate springing up every week, and waves of anxiety rolling in and rolling out. I felt a bit guilty sweating the question of whether my documentary project's 100-year hook was going to sail off into the sunset, considering many of the crew I had grown to know so well were sweating their livelihoods. Whether sailor, photographer, or garden variety steamboat geek, the twisted irony of No. 17 steaming until her 99¾ birthday then being lopped off into a headless, soulless barge–putting half the crew out of work–was a bitter pill to swallow. We waited, cussed violently, speculated endlessly and trolled for information that wasn't notoriously unreliable steamboat gossip.

Meantime, the price of diesel fuel spiked, and the boat continued to haul cement, very, very efficiently, as she has always done, spring into summer, fall into winter for the past 40 years.

Challenger burns unrefined Bunker C-type fuel oil, which became a significant cost advantage when the price of diesel shot up 51% in the latter half of 2005. This fuel spike–driven in part by a terrible hurricane season in the Gulf of Mexico–nearly bled me white, as I drove the wheels off the Honda during the "off-boat" phase of the photography. As fate would have it, the gas crunch that nearly shut me down may have helped save the boat. By the autumn of 2005 diesel had touched $3.15 or more per gallon, tug-barge conversions and their inherent limitations were being reassessed industry-wide, and *Challenger*, it seemed, was still very, very good at her job. When the bottom line was reckoned at the close of the season, her new owners in Toronto were favorably impressed, and talk of giving the boat the chop was shelved–at least until her next mandatory hull survey in 2009. She fit-out as usual in April 2006, freshly painted as had always been customary (a rare treat in an era when most lakeboats only get painted up during their five-year hull survey). After several warm-up laps in early April, barring an encounter with sea monsters or a UFO abduction, the magic moment was at hand.

Trip eight of the 2006 season found the *St. Marys Challenger* departing Charlevoix, Michigan, loaded with 9200 tons of Type I cement at 0930 on April 28, 2006–the centennial anniversary of her maiden voyage from the Great Lakes Engineering Works. Not surprisingly, there was little or no fanfare. If not for two days of rooting around the microfiche thickets at the Michigan History Museum in Lansing, I would have missed it myself. I packed along some de facto centennial cigars to share with Captain George Herdina, who I knew to be quite the stogie man. One of my favorite phenomena on the boat was watching the Old Man chomp a cigar and recite steering commands to his wheelsman as they manhandled the 552-foot ship up some unholy bullhead creek where no 552-foot ship had any business navigating. (All lake boat skippers are respectfully called "Old Man," though never to their face.) Ship handlers on the Inland Seas rank among the very best in the world, and Herdina, a 44-year veteran lake sailor, is the stoical master of his domain. Watching him creep the boat up the Kinnickinnic River where you could hear deadfalls on the riverbank snapping from the hull suction, or threading the huge boat through the Soo Line jack-knife bridge immediately after a 90-degree kink in the Manitowoc River in the ice, in the dark, was quietly thrilling. Like the archetypal airline pilot, the Old Man was imperturbable–no matter how far gone the game was. His demeanor, especially his demeanor on the open radio, never rose above bored courtesy. He had a way of keeping things loose with dour jokes insinuated between the steering commands that never failed to crack me up. If six torpedoes were coming at us, I'm pretty sure he would have make some dry remark and chewed his cigar only slightly faster.

On that less-than-diamond-studded afternoon in April, downbound off Big Sable Point, the Old Man and I smoked our centennial cigars out on the bridge wing and said little. The weather was fine. The boat delivered her split load to the Milwaukee and Ferrsyburg, Michigan terminals, returning to the plant at Charlevoix two days later, ready for another routine cargo–business as usual. In spite of my overripe historical posturing, it turned out to be the most ordinary of extraordinary days.

Centennial

For the record, the hundred-year milestone is a slippery thing, and the semantical hair-splitting is important to spell out. There are tugboats on the lakes that are well over 100-years old, a handful of them capable of getting up and going under their own power. Both the *E.M. Ford* and *J.B. Ford*, cement boats themselves, are older than *Challenger*. Launched in 1898 and 1904 respectively, both were relegated to duty as dockside storage zombies before either reached their diamond jubilees. In 2005 the sandsucker *John R. Emory*, which had laid at Erie, Pennsylvania for years, broke 100, and looked to have found a new lease on life with owners off-lakes, before the deal fell apart. There are steam yachts and museum pieces on both U.S. coasts and in Europe that are operational and older than No. 17. But nowhere in all the world will you find another large, (say for this purpose in excess of 10,000 gross tons) self-propelled carrier engaged in the routine delivery of commercial cargo at 100-plus years of age.

We can absolutely certify that she is the first (and presumably the last) lake boat to ever dip a toe south of the 100-year line. The next oldest vessel still active on the lakes is the M/V *Maumee*, (ex-*Calcite II*) at 78 years. The *Maumee* has been engaged in the poisonous but profitable trade of hauling road salt for nearly two decades, and the odds of her hull surviving another 22 seasons of ABS and Coast Guard scrutiny are virtually nil. *Challenger*'s accomplishment has never been (and I'm willing to bet my next two paychecks) will never *be* equaled in the history of world maritime commerce. Not even if that commerce goes on for another 5,000 years. Did I say that out loud?

All that you can't leave behind

In 2007, 101 years after her maiden voyage and 41 years after her conversion to a cement bulker, *Challenger* remains a floating conundrum. She is antique enough to qualify as the only laker that is still hand-steered at all times by a skilled wheelsman–no "Iron Mike" gyro helm that has been standard-issue on lake boats since World War II. She still has lignum vitae shaft bearings– lignum vitae being an exotic hardwood virtually extinct for decades–and she is the last long ship that still hoists anchor with a Smithsonian-caliber, reciprocating steam, anchor windlass engine.

The men who built her, rebuilt her, and the men who have maintained her through the decades can stand tall. In spite of these antediluvian characteristics, she remains one of the most efficient vessels in the business. That is her true pedigree. The shipping industry does not concern itself with historical preservation; at 101, *Challenger* and her crew still make money. At the end of the day, that is the most important standard by which a working steamboat is judged.

There will come a day when the *Challenger* can no longer raise steam. At the end of this decade or the next, or the next, she will sail away at last, having earned the peace of an endless winter slumber. Her hull may wander around in suspended animation for 30 more years, inelegantly butted around by a tug. But her long career as a Gideon's Warrior of the Steamboat Age will have run its course. The cold and snow will turn rivers and harbors to stone, and the infinite space in the middle of a continent will be empty of the silhouette of lake boats. The last thing the men and women who have made their homes aboard her will see as they leave with sea bags squared away is that extraordinary plume fading from her stack. Until that moment comes, she will roll on, a tiny island of souls under a cold sky, and an indomitable mechanical memory pushing on against the current of time and the inevitable. For now, she remains a defiant line on the horizon, steam bright in the air, a pure white banner flying out to rally the true believers against diminishment and the final goodbye.

The *St. Marys Challenger* arrived at the Chicago Seaway terminal with a lay-up load on December 18, 2006. The last entry in her logbook for the centennial season reads "finished with engine." The magic 100-year milestone had gone down to the lakebed without much of a splash. The pilothouse windows were shuttered up and deadwires run out against the winter wind. Officers and crew gathered their things together and made calls to arrange transportation to the airport. I shouldered my camera bag and thumped up and down the dock shaking hands with shipmates and thinking back to a solitary Christmas Eve downriver. Steamboat men are not a sentimental gang. I lingered anyway, as darkness swallowed up the forward end. The whistle was blown down in a dying arc that echoed over the silos and salvage yards and the sense that something very special had come to end hung in the air over Lake Calumet.

The last of the forward end crew caught cabs or met wives and family dockside, anxious to get underway on the trip home. I watched the boat recede in my rearview mirror, cold rain grey on the windshield. I hoped I got it right. It occurred to me, as I merged into the mundane stream of travelers on the interstate, that I'd spent the last years of my youth cutting roast beef with a screwdriver. It's the Lakes you come to love… the Lakes that inspire, that elevate sailing or photography or hauling cement up and down the coast to the status of a calling. Without them, everything contained in this book would just be dead machinery.

The after end gang spent a week or so putting the engine to rest and making ready for another season. *Challenger* would raise steam again in the springtime, 101 years after tasting sweet water for the first time. Her crew would return for another chaotic fit-out week in April 2007. I would not be among them, having outlived my useful reason to ride out with the boat.

I turned my mind to the building of this book, always craning over the computer for a glimpse of a lake boat on the horizon, and surrounding myself with photographs, stacked up three and four deep in my studio, procrastinating for six months, hesitant to call the project finished. They are flat things in the end, photographs, bereft of the smells of diesel fuel and fresh cinnamon in the galley and the laughter of friends. I've filled these pages with photographs to remind myself, trying to hold on to a feeling that has already fallen astern. The *St. Marys Challenger* steams on, her lights fading away over the line that separates sky from water, and I am left with photographs and the bittersweet ache of having lived too long with a single dream.

Hull No. 17, Spring 1906
The S.S. *William P. Snyder* lies in the Great Lakes Engineering Works' fitting-out berth at Ecorse, Michigan. The hull of steamer *James Laughlin* nears completion on the builder's stocks to the right.

LAUNCH OF STEAMER WILLIAM P. SNYDER.

Another link has been made in the great chain which is being forged to connect the iron making industry of the country with the sources of its raw material. One of the characteristic features of the past two or three years has been the distribution of the great steel making companies to control the avenues of transportation leading from their mines to their furnaces. This policy has resulted in the placing of a number of orders for freighters with lake ship yards. The latest vessel of this class to be launched was the Wm. P. Snyder which went overboard at the Ecorse yard of the Great Lakes Engineering Works of Detroit on Saturday last. This steamer is building for the Shenango Steamship Co. of Cleveland and is intended to carry the ore of the Shenango, Webb and Whiteside mines owned and operated by the Shenango Furnace Co. This new steamer is one of the largest on the lakes, being 550 ft. over all, 530 ft. keel, 56 ft. beam and 31 ft. deep. She has thirty-three hatches, spaced 12 ft. centers. Her engine is triple expansion with cylinders 23, 37 and 63 in. diameters by 42 in. stroke, supplied with steam from two Scotch boilers 15 ft. by 12 ft., with induced draft and allowed 175 lbs. pressure. The steamer will carry 10,000 gross tons of ore.

The launch was successful in every way. The launching party left on a special car and had luncheon enroute to the ship yard. The steamer was christened by Miss Mary B. Dyer, of Pittsburg, daughter of Mr. C. D. Dyer, secretary of the Shenango Steamship Co. Mr. Wm. P. Snyder after whom the vessel is named, was unable to be present as he is in the south with his family. Those in the launching party included the following: From Pittsburg, Charles H. McKee; C. H. Hayes, vice-president Pittsburg Trust Co.; Henry Irwin Jr., treasurer Shenango Furnace Co.; C. D. Dyer, vice president Shenango Furnace Co.; W. A. Barrows, Jr., general manager Shenango Furnace Co.; A. W. Renwick, O. P. Scaife, Jr., S. C. Irwin, Mrs. Dyer, Mrs. Renwick, Mrs. Barrows, Mrs. Irwin, Mrs. H. D. Rankin, Miss Christine Beale, Miss Mary B. Dyer. From Cleveland: R. P. Ranney, W. B. Davock, H. M. Herriman, Ralph D. Mitchell, Capt. Peterson, Mrs. Ranney, Mrs. Davock. From Detroit: George H. Russel, H. C. Potter, Jr., H. P. Davock, L. C. Waldo, Alexander K. Gage, Harlow N. Davock, Clarence W. Davock, Antonio C. Pessano, John R. Russel, Mrs. Russel, Mrs. Davock, Mrs. Waldo and daughters, Mary and Louise, Mrs. Gage. From Ypsilanti: Miss Nan Olmstead. Prof. H. C. Sadler of the University of Michigan embraced the opportunity to attend the launch with about thirty students in his class in marine engineering and naval architecture.

Money will not be spared to make the Snyder in respect to interior furnishings one of the most elegant freighters on the lakes, Mr. Snyder having personally made a donation for this purpose.

MR. W. B. DAVOCK, MANAGER OF THE SHENANGO STEAMSHIP CO., MISS MARY B. DYER AND MR. C. D. DYER.

MISS ANNIE RUSSEL AND MR. JOHN A. UBSDELL.
Query after the launch. What was it Miss Russel said to Mr. Ubsdell?

STEAMER WM. P. SNYDER.

Built by the Great Lakes Engineering Works of Detroit, Michigan, and equipped with Kidd's Patent Hawse Pipe which allows of the anchors being very snugly stowed, bringing them in flush with outside of the hull as shown in above picture.

Above and opposite: Excerpts from *The Marine Review* **of 1906**

CAPT. HENRY PETERSON.

Broadside perspective of the *Snyder* moored in Cleveland Harbor

THE DETROIT FREE PRESS: SUNDAY, APRIL 29, 1906.

MARINE NEWS

Snyder on Maiden Trip.

The new steamer William P. Snyder left the Ecorse yard of the Great Lakes Engineering Works early yesterday morning on her maiden trip. She is bound for Duluth for ore. Capt. Henry Peterson is in command of the vessel.

The *William P. Snyder's* observation lounge, circa 1917

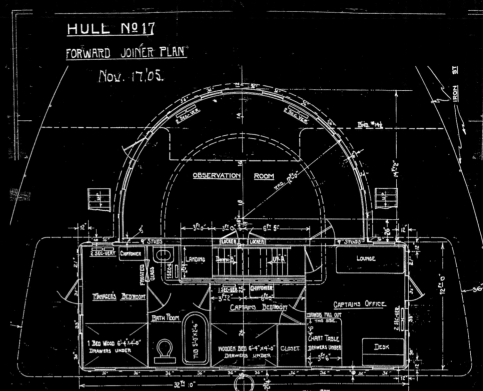

Above: Joiner plan of the *Snyder*'s Texas Deck
Below: Photo circa 1917, captioned
"Mr. Snyder's Private Room"

Above: A.E. Young's camera catches the five vessels of the Shenango fleet upbound at Sault Ste. Marie, Michigan, 1923

Opposite: View of the engine room operating deck and the *Snyder*'s original 1600 H.P. triple-expansion steam engine, circa 1917

The *William P. Snyder* sliding the Poe Lock approach, upbound at Sault Ste. Marie, Michigan, circa 1915

Ship and 30 Men Lost in Storm

May Be Victim of Heavy Storm on Upper Lakes

LAKE COAL SHIP WITH 30 MEN MISSING

Believed Caught in Heavy Storm; Overdue 28 Hours Here

A heavy snowstorm and high gale, sweeping over the Northern Wisconsin and upper lakes district, was believed today to have caught in its grip the Milwaukee bound steamer, William P. Snyder, carrying a crew of thirty men and 11,500 tons of coal.

The steamer, which left Ashtabula, O., and was reported last off Mackinaw, Mich., on Thursday morning, was thirty-one hours overdue at 3 p. m. today. The storm broke immediately after its last report and no word has been received from the craft since, despite continued efforts to locate it.

The craft is five hundred thirty feet long and one of the largest on the lakes. It is of steel and owned by the Shenango Steamship Co. of Cleveland. J. J. Slade of Detroit, is captain. Its cargo is consigned to the Milwaukee Western Fuel Co.

Fear was expressed here today that the collier, William P. Snyder, more than a day overdue in Milwaukee on a trip from Ashtabula, O., had been caught in a severe storm raging over the northern lakes district. The vessel, shown here, is a five hundred thirty foot craft, carrying a crew of thirty men and 11,500 tons of coal.

MOST SEVERE OF YEAR.

The snowstorm in which the William P. Snyder is believed to have been caught is the most severe of the year, said reports from up the lake, and has demoralized upbound traffic north of Sault Ste. Marie, where the harbor is cluttered with vessels fearing to venture into the heavy weather. In some quarters belief was expressed that the William P. Snyder has anchored off some remote port until the storm abates.

FREEZING AT SUPERIOR.

At Superior, the lowest temperature yesterday was 32. Colder weather was predicted for that district.

Milwauke today experienced the coldest weather of the season, the mercury registering 36 at 7 and 8 a. m.

Heavy frosts and continued cold are predicted for tonight in the Chicago and Milwaukee districts, with slightly warmer weather early tomorrow in northwest Wisconsin.

10-20-1923

Above: J.J. Slade of Detroit, the *Snyder*'s captain, 1923

October 1950 With repowering work complete, Christy Corporation engineers and the *Elton Hoyt II's* captain hoist a broom up the forward mast, the traditional emblem of a successful sea trial.

7½ years' cargo in 6 years for the "Alex D. Chisholm"

Since it was repowered with a  SKINNER MARINE UNAFLOW STEAM ENGINE

THE 10,000 ton bulk carrier, "Alex D. Chisholm," owned by The Interlake Steamship Company, Cleveland, Ohio, was repowered in 1950 with a 3,000-hp. Skinner Marine Unaflow Steam Engine.

The record for 1955, the sixth operating season with a Skinner Engine, compared with performance before repowering shows that:

Power was 50% greater, with less fuel consumed per mile. 440,058 tons of cargo were carried, an increase of about 25%. 35 round trips for a total of 52,706 miles were made, *completing 6 seasons without interruption.*

The "Alex D. Chisholm" is one of many Great Lakes ships which have been repowered with Skinner Marine Unaflow Steam Engines, for improved performance and lengthened service.

Skinner Marine Unaflow Steam Engines are available in capacities from 400 to 6000 horsepower, with from two to six cylinders, in various lengths of stroke, and a wide range of speeds.

For Over 80 Years, Doing One Thing Well—Building Steam Engines

SKINNER ENGINE COMPANY, ERIE, PA.

Foreign Licensees
Canadian Vickers Limited, Montreal, Canada • Amsterdam Drydock Cy., Amsterdam, Netherlands
Ateliers & Chantiers, Le Trait, France

Left: Upbound on the Detroit River abreast of Zug Island

Right: Crossing Lake St. Clair as the *Alex D. Chisholm*

Below: Spring fit-out as the *Elton Hoyt II*

Above: The *Alex D. Chisholm* downbound at Detroit in August, 1955

Opposite: Hull No. 17, converted to haul bulk cement, is floated off the drydock at the Manitowoc Shipbuilding Company, early summer, 1967

Inset: Billboards and bunting welcome the newly christened *Medusa Challenger* to the Cleveland waterfront June 27, 1967

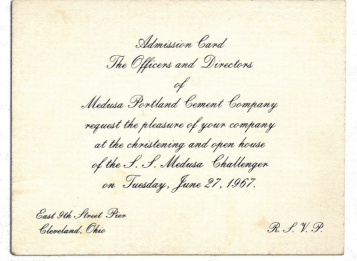

Admission Card
The Officers and Directors
of
Medusa Portland Cement Company
request the pleasure of your company
at the christening and open house
of the S. S. Medusa Challenger
on Tuesday, June 27, 1967.

East 9th Street Pier
Cleveland, Ohio   *R. S. V. P.*

Smashing Launching

SMASHEROO—Irene Sedgwick, sponsor of the Medusa Challenger, set off a deafening round of whistles from vessels along the lakefront yesterday with this mighty swing. The christening of the 552-foot ship was held at the foot of E. 9th Street.
Plain Dealer Photo (Dudley Brumbach)

Above right: Admission card from *Challenger*'s re-christening ceremony at Cleveland, Ohio

Below right: A champagne crack on the nose grants the 60-year-old boat a new lease on life

Above: The Chicago River "jinx ship" inbound at State Street, summer 1970

SHIP CASTS ITS SPELL ON CITY BRIDGES AGAIN

BY WILLIAM CURRIE

The Medusa-Challenger, nemesis of the Chicago river bridge tenders, struck again last night. This time the ship cast its spell on three bridges.

Last month, the 562-long cement ship out of Detroit set a record when it tried to get thru the river three times and each time a bridge or two stuck.

The ship arrived at the river at 6 p. m. yesterday.

Luck Runs Out

The Outer drive bridge was no problem. The Michigan avenue bridge worked. Clear sailing ahead.

Luck ran out at 7 p. m., at the Wabash avenue bridge. After the Medusa-Challenger got thru, Edward Windle, 30-year veteran bridge tender, threw his switches. But the south leaf wouldn't drop. There was a power failure.

The Dearborn and Clark street bridges survived the jinx. And the La Salle street bridge worked well.

Luck ran out again at the Wells street bridge as the Medusa-Challenger approached. The bridge wouldn't budge. The problem was compounded because the ship is by 100 feet the longest vessel to use the river.

Three Bridges Out

With the ship's prow not quite to the Wells street bridge, the stern was under the La Salle bridge.

At 7:20 p. m., the Wabash avenue, the La Salle street, and Wells street bridges were out of commission.

Starting with Wells street problem, city electricians went to work and by 8:30 p. m. Medusa-Challenger sailed again.

Week-end traffic was light and no traffic problems occurred. But the Medusa-Challenger will be coming back— all summer long.

CEMENT Transit Co.
Subsidiary of
MEDUSA PORTLAND CEMENT COMPANY

S.S. MEDUSA CHALLENGER

CAPTAIN		CHIEF ENGINEER
JOHN C BRADLEY	THANKSGIVING-MENU	ROY H KLEIN
	NOVEMBER-25-1976	

PREPARED-BY
LARRY GEIGER STEWARD
HARLEY THIES 2nd COOK

APPETIZERS

JUMBO   CHILLED
SHRIMP COCKTAIL   FRUIT COCKTAIL

ROS'E WINE - APPLE CIDER
ASSORTED OLIVES PICKLES & NUT CUPS

ENTREE

LONG ISLAND DUCKLING
ROAST TURKEY   CORNISH GAME HENS
ALL WITH DRESSING
BAKED VIRGINIA HAM & RAISIN SAUCE

VEGETABLES

WHIPPED POTATOES   CANDIED SWEET POTATOES
ASPARAGUS SPEARS WITH CHEESE SAUCE
BUTTERED SQUASH   MASHED BUTTERED RUTABAGAS

SALADS

FRUIT SALAD-COLD SLAW-CARROT PINEAPPLE SALAD
STRAWBERRYS and GELATINE SALAD
JELLIED & WHOLE CRANBERRY SAUCE

HOT DINNER ROLLS

DESSERTS

MINCE PIE   PUMPKIN PIE   APPLE PIE ALL ALAMODE
HOLIDAY ICE CREAM   SHERBETS

REFRESHMENTS

COFFEE TEA MILK HOT CHOCOLATE SOFT DRINKS BEER
ASSORTED CANDY & NUTS GUM CIGARS CIGARETTES

Larry Geiger 0633 Collection — #0633

Bob Wiedrich
Agony, despair, and a lake rescue

"YOU REACH A point where you think it would be better to let go of the rope and get it over with."

Those are the words of John S. Findlay, a 51-year-old consulting engineer, who survived 15 hours in the waters of Lake Michigan a week ago clinging to an overturned powerboat buffeted by several storms.

"The flotation material in the three hulls of the boat was getting so waterlogged the keel was just a few inches above water. We were sinking. We were going down.

"I figured that in another 15 or 20 minutes we were through. We would not survive.

"DAWN HAD COME slowly with an ugly gray cast. Visibility was bad. The waves were still running about three to four feet high. We were sitting so low in the water we figured nobody would be able to spot us from a passing ship.

"And we were so exhausted, so paralyzed from clinging to the capsized boat

> *"A freak wave cresting like a wall of water hit us on the starboard side. It must have dumped a ton of water into the bow and stern."*

for so many hours, that we knew we would be unable to wave our arms.

"Then out of the mist I saw something that looked like a building. For a minute, I thought I was hallucinating.

"Then I realized it was a big freighter and that we were drifting more or less into its path. It was riding so high in the water without cargo that I was afraid no one on deck would be able to see us.

"But suddenly I found myself waving frantically. Somehow I mustered strength I didn't know I had left.

"And then I heard the most wonderful sound in the world — the ship's engines stopping 200 or 300 yards away. And I knew we were safe, that we had made it."

THE MEDUSA Challenger—the 562-foot-long cement carrier Chicago bridgetenders call their jinx ship—had arrived just in time.

Capt. Jack Bradley, the skipper, and the men on the bridge had spotted a bright orange lifejacket in the water as they sailed outbound from Chicago for Charlevoix, Mich.

And in moments after spotting them, the crewmen lowered a special rescue craft to pluck Findlay and a companion, 29-year-old James Mueller of Glenview, from the still storm-tossed waters of Lake Michigan.

During the ordeal, another companion, Fred Gerlitz, a 68-year-old retired carpenter, also from Glenview, had been swept away and presumed drowned after a massive wave tore all three men from the boat.

That had been about 8:30 p.m. the previous night, some 4½ hours after the trio had first been hurled into the water by another giant wave perhaps seven to eight feet high.

"A little while later, the waves started running about four feet high. The rain began. The winds were gusting 25 miles an hour or more.

"Then a freak wave cresting like a wall of water hit us on the starboard side. It must have dumped a ton of water into the bow and stern. And the boat just went down with the weight of all that water.

"Fred Gerlitz grabbed for a pair of lifejackets and handed one to Mueller. I grabbed for the radio to send a Mayday signal for help. But the damned thing had shorted out in the water.

"That was when the boat rolled over. I didn't have a lifejacket. I found myself trapped underneath. I nearly panicked. To this day I can't tell you how I got out.

"All I know is that I surfaced near the bow. Mueller was beside me. Fred was at the stern."

THAT MUST have been soon after 4 p.m., shortly after the Coast Guard reports having received Findlay's request for weather information.

Gerlitz tied a piece of rope around his waist and secured it to the overturned boat and held onto the engine. Mueller and Findlay held on to another length of rope which they tied to a towing eyelet at the bow.

"When I first got clear, Fred obviously had been trying to get his lifejacket on and it drifted away from him.

"At first, we attempted to climb onto the capsized boat and pull her right side up. But the stern was facing the waves. And the fiber glass was smooth and very hard to hold on to.

"Jim Mueller and I tried to hang onto the belly of the boat, but the waves kept knocking us off.

"Jim's watch was still working. So I know that about four hours later an exceptionally big wave hit us. I never saw it coming. It tore all three of us loose.

"By the time Jim and I got back to the bow, I got a glimpse of Fred in the water and then he disappeared.

"THE STORM lasted all night. I kept praying it would abate so we could get some rest.

"We finally managed to crawl onto the hull back at the stern. I held onto the rope Fred had tied around his waist before he was washed away and Jim held onto my belt.

"That's how it was for the next 11 hours as the waves and the winds and the rain lashed at us until our hands and legs became so paralyzed we weren't sure we even were holding onto anything any more.

"Jim's watch stopped. But I figure about 4 o'clock in the morning, the beam of a Coast Guard helicopter's light passed across the surface of the water maybe 100 feet away from us.

"I could hear the helicopter. But the crew couldn't see us. We had no flares. We had nothing but rough seas.

"It's funny what happens to you during an ordeal like that. At one point, I thought I was smoking a cigaret. I could actually taste it.

"Then another time I thought I saw the masthead lights of a sailing ship, which of course was impossible in that storm."

THE VOYAGE had started pleasantly enough almost 24 hours before the Medusa Challenger's last-minute rescue when the three men set sail from a boat launching ramp at the foot of Church Street in Evanston in search of coho and chinook salmon.

Findlay, who lives in Wilmette, was proud of his 18-foot-long, tri-hulled, fiber glass powerboat which was very stable and hard to sink.

He headed the craft north toward Highland Park and, when the men found the fish scarce close to shore, steered southeasterly into the open lake to a favorite spot 7½ miles east of Wilmette.

The weather was excellent. But again, the fishing was bad. So Findlay moved the craft to a point 11 miles offshore before Mueller finally snared the sole catch of the day.

"WE WERE STARTING to come back in because the fish weren't biting when I noticed the shoreline didn't look too healthy," Findlay said. "It was clouding over and it looked like rain.

"So I called the Coast Guard on the radio for weather information and they told me that a band of thunderstorms was due to move over the lake, but that it probably would miss us.

AT 6:40 A.M. LAST Friday, Aug. 5, the Medusa Challenger sailed out of a churning sea to rescue the two survivors from waters 250 feet deep at a point 16 miles offshore and 18.2 miles north of the Wilmette Coast Guard station.

"When we were helped on board, I told the captain his must be a lucky ship," Findlay said. "And Capt. Bradley replied with a grin, 'Did you notice the ship's name?'

"I said yes and I told him I didn't care if the Medusa Challenger jinxed all of Chicago's bridges for the rest of her life."

The cement carrier, as you probably recall, has an unfortunate reputation for prompting bridges over the Chicago River to get stuck in a raised position whenever she sails through.

"The crew dried us off and gave us clothing and the medical man even gave me a shot of brandy in my coffee," said Findlay. "Jim Mueller wouldn't touch it. He was as pale as a ghost. And so was I.

"But now that we've recovered, we are going to go down to the docks the next time the Medusa Challenger is in Chicago and personally thank Capt. Bradley and his crew.

"They were magnificent. And so is their ship."

Opposite left: *Challenger's* long-time Steward Larry Geiger chats with the port chaplain in the ship's galley, circa 1974

Opposite right: Menu from Thanksgiving dinner, 1976

Opposite below: Postcard of the *Medusa Challenger* in bicentennial livery

Right: Snapshots from a 1977 rescue detailed in the above Chicago Tribune item, and the rescue of a sinking cabin cruiser in 1975

Ports of Call

The Medusa Challenger makes regular stops at all of Medusa's facilities. It is Medusa's ability to serve this important geographical region with timely, efficient delivery that keeps Medusa the leading force in cement and masonry products.

From our modern cement plant in Charlevoix, Michigan, the Medusa Challenger sails to:
- Cleveland, Ohio in 44 hours
- Toledo, Ohio in 37 hours
- Detroit, Michigan in 31 hours
- Manitowoc, Wisconsin in 13 hours
- Milwaukee, Wisconsin in 19 hours
- Chicago, Illinois in 27 hours
- Ferrysburg, Michigan in 17 hours

Our Manitowoc terminal serves land based Wisconsin silos in Green Bay, Rhinelander, and Stevens Point.

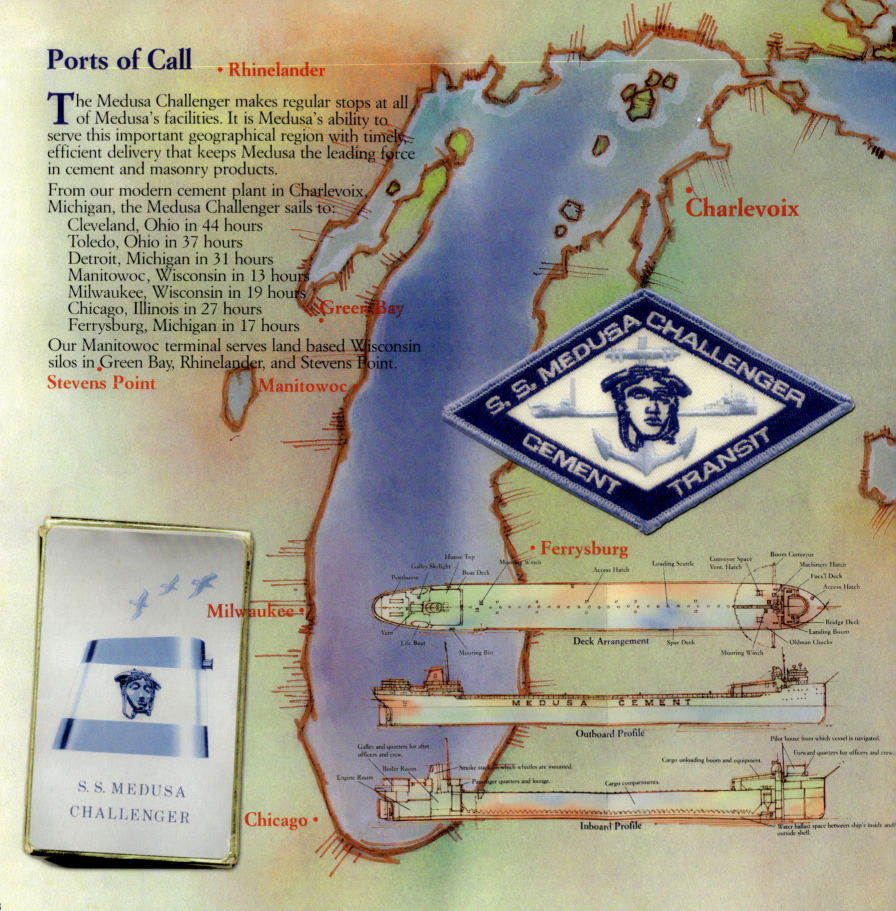

The Great Lakes

The range of the Medusa Challenger covers most of the Great Lakes. The basins of the Great Lakes were scooped out by the southward movement of glaciers during the Ice Age. Today the five lakes: Superior, Michigan, Huron, Erie, and Ontario, form the greatest connected area of fresh water on earth. If put together, the lakes would more than cover the states of New York and Pennsylvania.

The Challenger and our new 8,200 ton barge are two of the most advanced and efficient modes of cement transport in the industry. Our vessels reflect Medusa's commitment to meeting the supply requirements of our customers on the Great Lakes.

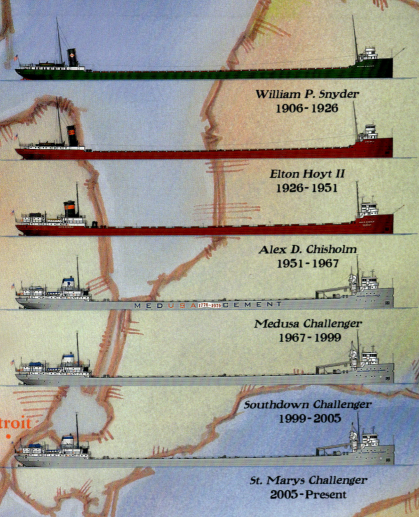

William P. Snyder
1906 - 1926

Elton Hoyt II
1926 - 1951

Alex D. Chisholm
1951 - 1967

Medusa Challenger
1967 - 1999

Southdown Challenger
1999 - 2005

St. Marys Challenger
2005 - Present

Interesting facts about the S.S. Medusa Challenger

The length of the Challenger is 552′ 1″.

Her beam, or width, is 56′.

Her draft, or the depth of her keel below the surface, is 21′2¾″.

Total displacement, midsummer is 17,720 tons.

Fully loaded, she will move along at 14 miles per hour.

Empty, her speed increases to 14½ miles per hour.

Her propeller is taller than the combined height of two men, 14′3″, and weighs more than six automobiles, 13,000 lbs.

It is estimated that she will haul approximately 950,000 tons of cement annually in her eight holds.

One full load of cement, 11,600 tons, is sufficient to build a two lane highway stretching over 10 miles in length.

One full load of cement will also make 6,700,000 concrete blocks.

Owner: Cement Transit Company (a subsidiary of Medusa Portland Cement Company).

Operator: Cement Transit Company

Christened: June 27, 1967

Nautical Terms

DRAFT – depth of vessel from water line to keel or bottom of vessel.

BULKHEAD – wall or partition aboard ship.

AIRPORT – window.

HOLD – cargo space.

PORT SIDE – left side of vessel, when looking forward.

STARBOARD SIDE – right side of vessel.

GALLEY – kitchen.

Athwartships	right angle to keel line
Fore-and-Aft	parallel to the keel
Windward	toward the wind
Leeward	away from the wind
Beam	breadth of ship
Hawser	heavy rope
Binnacle	protective casing for compasses
Corinthian	amateur sailor

Above: The *Southdown Challenger* arrives at Bay Shipbuilding's Berth 15 for her five-year survey
Opposite: Shipyard rivet gang buttoning up hull plates at the turn of *Challenger*'s bilge, July 2003

51

Centennial

Southdown Challenger
2003-2004

Above: Five-year hull survey, on the drydock at Bay Shipbuilding
Opposite: View of the boat's lapped and riveted counter stern
July 2003

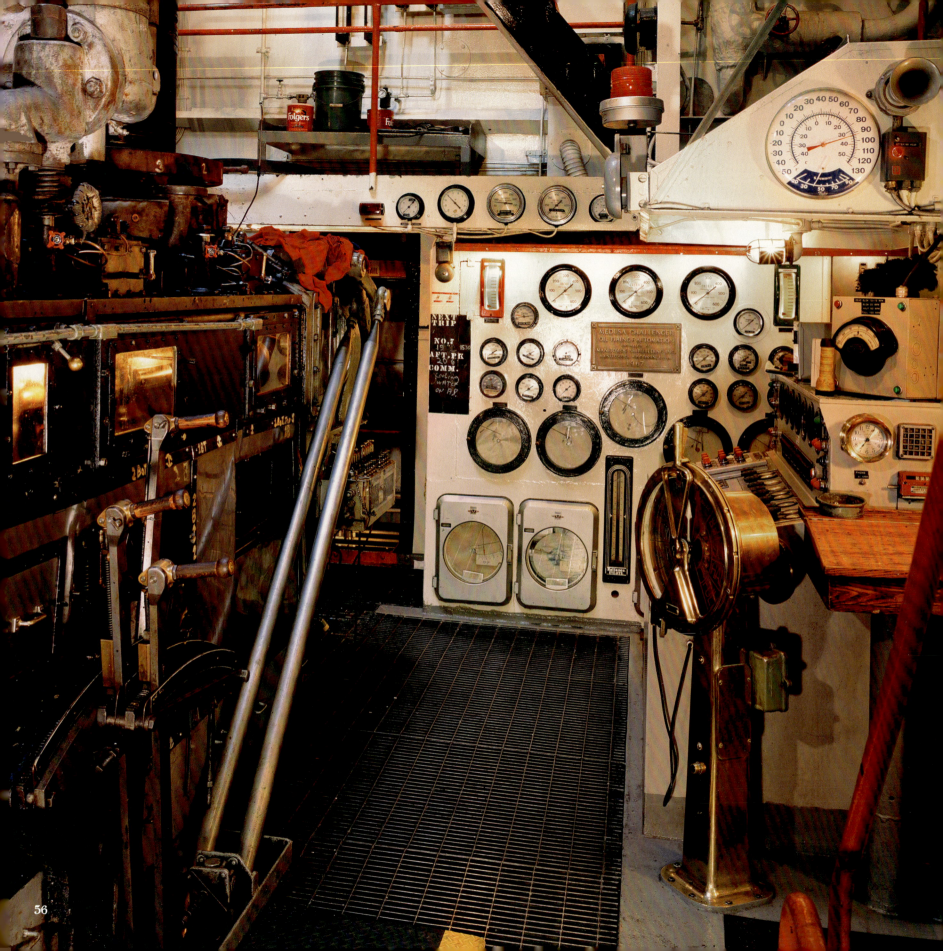

Above: Skinner Unaflow steam engine with piston one pulled for repair
Opposite: View of the engine room control station and telegraph
July 2003

Below: *Southdown Challenger*, bow view, on approach to the loading slip at Charlevoix, Michigan
Opposite: Passing the breakwater hulls of the *Charles S. Hebard* (1906) and *Amasa Stone* (1905) August 2003

Challenger's charmed life is showcased every time she arrives home to the cement plant at Charlevoix, Michigan. On approach to the loading slip there, she slides past the dry bones of the lakers *Charles S. Hebard* and *Amasa Stone*. Both boats were launched around the same time as Hull No. 17, and both were retired from service over 40 years ago. They were filled with concrete and sunk as ready-made breakwalls when the Medusa Cement Company built the manufacturing facility in 1966.

Above: 2nd Assistant Engineer Al Wirgau, 12-4 engine room watch August 2003

Above: Approaching the *Ryerson* Cut on the Manitowoc River August 2003

Above: Downbound along Refinery Row on the upper St. Clair River
Opposite: View of the Milwaukee terminal from Skipper Bud's Marina
October 2003

Above: Passing the ruins of the ACME steel works at South Chicago
Opposite: Outbound on the Calumet River, rounding Wisconsin Steel Bend
November 2003

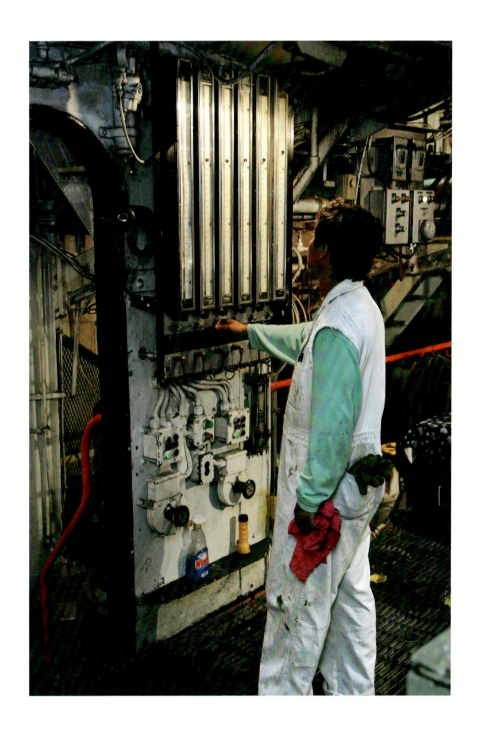

Above: Oiler checking ballast water levels via King-Gauges
Opposite: Wheelsman's post, downbound on Lake Erie
November 2003

Above: Sunset on Lake Michigan November 2003

Arriving in ports at all hours, lake sailors rarely depend on shoreside line handlers to assist in tying up the boat. Instead, a pair of deckhands are swung over the side on the Bosun's chair and landed on the dock. The landing party is thrown heaving lines secured to 2" diameter steel cables which are hauled in and looped over mooring spiles... quite an adventure if it's blowing a screecher and the dock is icy.

Above: Pulling cable to mooring spiles at Charlevoix
Opposite: Riding the Bosun's chair during docking operations
December 2003

Above: A frigid shift at South Chicago
Opposite: Slogging through pancake ice on Lake Huron
December 2003

Above: New Year's Eve at South Chicago, passing the *John D. Leitch*
Opposite: Tanker vessel *William L. Warner* arrives on Lake Calumet
January 2004

Above: Towing out of winter lay-up at the Milwaukee terminal
Opposite: Work gang removing pilothouse shutters at spring fit-out
April 2004

There is no more apparent sign that a lake boat is about to wake from her winter slumber than removal of the pilothouse window shutters. Like the return of the robin or the break-up of river ice, fit-out of the lake fleet in April once qualified as an authentic rite of spring in many harborside communities.

Above: View from the unloading conveyor at Milwaukee, Wisconsin
Opposite: Builder's trademark in the boat's windlass room
April 2004

Above: Making up the tow post
Opposite: Tug *Arkansas* assists on the Kinnickinnic River
April 2004

Above: Deckhand Bonita Vineyard, Able-Bodied Sailor
Opposite: Westbound in the Straits of Mackinac
April 2004

Bonita Vineyard was a rare sight on the lakes. Daughter of a career Lake Michigan car-ferry sailor, she shipped out after high school and was one of the *Challenger*'s most experienced deckhands. Tough and very competent, she was usually first to go over in the Bosun's chair, and always ready to take the boys' money at long-haul poker.

Below: Unloading a split load of Type II at Ferrysburg, Michigan
Opposite, clockwise: Logging engine orders while manuvering, a coffee mug caddy in the pilothouse, sunset on Lake Huron, and porthole detail, aft
May 2004

This portrait of country music legend Willy Nelson hung proudly in 2nd Cook Joel Markle's porthole until kidnapped by masked banditos—possibly in retaliation for some unpleasantness up the street involving a case of malt liquor and a box of chocolate laxatives (it was hard to tell, exactly). Our beloved Uncle Jo-Jo vowed to pursue his heart-rending "Free Willy!" sloganeering and profanity-laced castigation of the perpetrators' mothers and ancestors until the picture was returned. To this day, the Red-Headed Stranger's whereabouts remain a mystery.

Above: Fuel trucks at Charlevoix
Left: 12-4 watch on the bridge wing
Opposite: Chief Engineer Mike Laturi answering engine telegraph commands
May 2004

Above: Departing Detroit, downbound, below the Ambassador Bridge
Opposite: Spring passage, upbound on the Detroit River
May 2004

Above: Sunrise, inbound on the Cuyahoga River at Cleveland, Ohio
Opposite above: A full load of Type 1 for the Cemex terminal at Toledo, Ohio
Opposite below: M/V *Fred R. White Jr.* squeeks past at Collision Bend
May 2004

Below: Passing straightdecker *Montrealais* in the St. Clair Flats
Opposite: Approaching the Blue Water Bridge at Port Huron, Michigan
May 2004

Below: Arriving in Milwaukee harbor, steamer *Alpena* departing
Opposite: Dockside view of *Challenger's* unique steam plume
May 2004

Behold, *Challenger*'s signature auxiliary steam plume. There is no more magnificent a sight on the wide waters of the Inland Seas. Low pressure steam is bled off the boilers to power ancillary equipment—winches, generators, etc. After doing its work, this steam is exhausted overboard, creating a unique calling card anywhere on the horizon and a fantastic photographic companion.

Above: First Mate Ray "Rocky" Groh navigating through the Livingston Channel
Opposite: Fireworks over Windsor, Ontario; unloading Type II at the Detroit silo
June 2004

This series was inspired by Joseph Turner's painting *Snowstorm* of 1842. It's also a testament to the hammer toughness of the Canon EOS system. Within minutes of the storm breaking I was soaked to the point that I might as well have gone overboard with the equipment rig in tow. I hung the camera body, flash unit, and lens in the warmth of the engine room overnight, and proceeded to blast away without incident the next morning.

Above and opposite: **Stormy passage at Port Huron**
July 2004

Below: Backing down the Cuyahoga River at Cleveland
Opposite: Dinner for VIP guests in the officer's dining room
July 2004

Above: Crew change at Detroit via mailboat *J.W. Westcott II*
July 2004

Above left: Glen's Market loading provisions at Charlevoix
Above right: Sunset on Lake Erie, eastbound for Toledo, Ohio
July 2004

Above: Oiler Andy Egressy makes rounds on the 8-12 engine room watch
Opposite: Engineer shifting the Unaflow, inbound on the Maumee River
July 2004

Below: Cleveland Harbor Light, inbound
Opposite left: Passing steamer *Arthur M. Anderson*
Opposite right South tower of the Mackinac Bridge
July 2004

The mailboat *J.W. Westcott II* is a unique fixture on the Great Lakes, and the only vessel in the world to have her own postal ZIP Code. The Westcott Company has been ferrying mail, sundries and personnel out to passing freighters on the Detroit River since 1874. Begun with a fleet of rowboats, Westcott's 'round-the-clock operation now makes as many as 6,000 transfers in a nine month season.

Above: Clearing the Veterans Memorial Bridge on the Cuyahoga River
Opposite: "Mail by the pail" from the *J.W. Westcott II* at Detroit
July 2004

Above: A full load of Type I for the Cleveland terminal
Opposite above: Passing steamer *Saginaw* at the CSX coal dock, Toledo
Opposite below: View from the helm stand, downbound at Belle Isle
September 2004

Above: Approaching the Ambassador Bridge, downbound
Opposite above: Arriving at the Grand Haven, Michigan piers
Opposite below: Passing the M/V *Canadian Miner* on the St. Clair River
October 2004

The crew found Miss Lucy abandoned on the riverfront at Cleveland. She adjusted to shipboard life very well, except when the anchor dropped. The din of the anchor chain thundering out would send her into the maze of pipes in the dunnage room overhead for days at a stretch. The Bosun took her home to Tennessee at the end of the season, where a routine check-up established that Miss Lucy was not in fact a miss.

Above: Unloading at the Miller Paving silo at Owen Sound, Ontario
Opposite: Miss Lucy, ship's cat
November 2004

Above: A "triple split" for the Cleveland, Toledo and Detroit silos
Opposite: Early cold at Charlevoix, loading Type II for Milwaukee
November 2004

Thanksgiving on a steamboat is traditionally an epic feast. The crew dresses up a bit and does their best to act as festive as possible. Some sailors will tell you the more the galley department tries to make the day feel just like home, the more they make everybody miss it. Serving up Turkey Day 2004 was a bit dicey, due to building weather on northern Lake Michigan. Let me assure you there is nothing more tragic than the meeting of a fresh apple pie with a cold, steel deck.

Above and opposite: Dawn in the galley, Thanksgiving morning
November 2004

119

Above and opposite: An old-fashioned steamboat Thanksgiving
November 2004

Above: Bow watch, downbound at Port Huron
Opposite left: Clearing the Blue Water Bridge, upbound
Opposite right: Passing the M/V *Algosteel* in the Huron Cut
December 2004

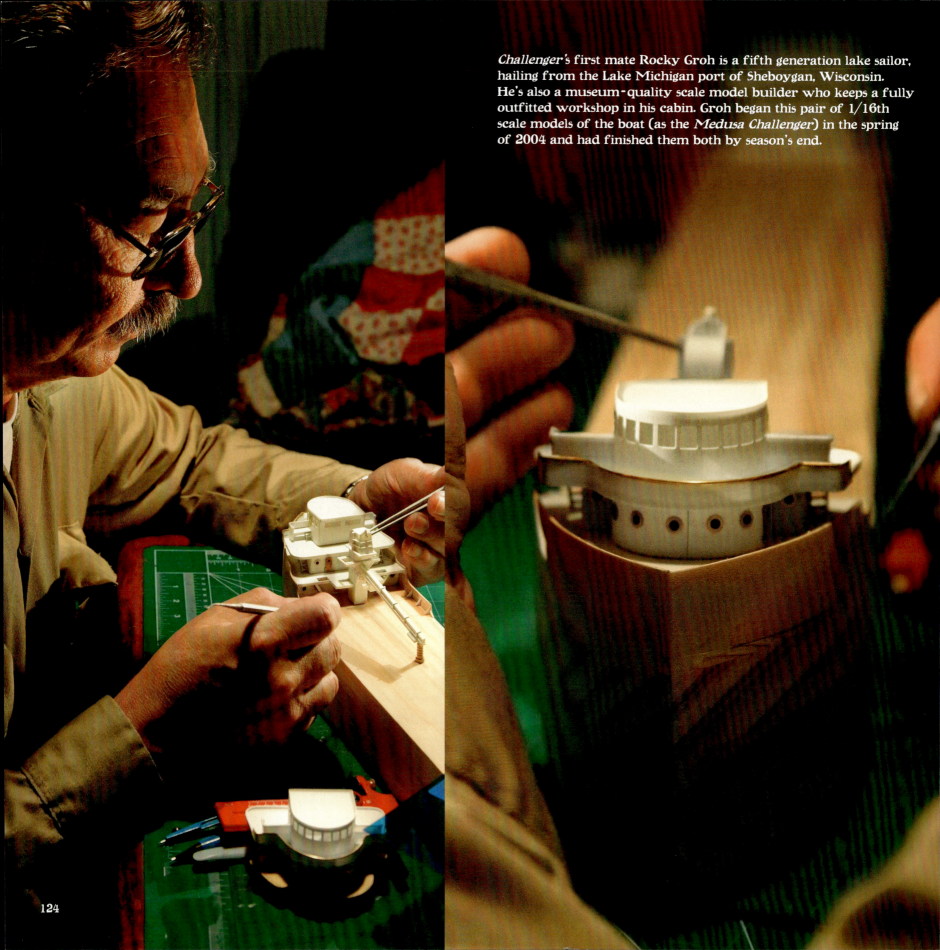

Challenger's first mate Rocky Groh is a fifth generation lake sailor, hailing from the Lake Michigan port of Sheboygan, Wisconsin. He's also a museum-quality scale model builder who keeps a fully outfitted workshop in his cabin. Groh began this pair of 1/16th scale models of the boat (as the *Medusa Challenger*) in the spring of 2004 and had finished them both by season's end.

Above: *Medusa Challenger* model, captain's office
Opposite: Mate's cabin, 1/16th scale model in progress
April/December 2004

Above: Ice-glazed unloading snorkel
Opposite: Deck view, morning after a sub-zero blow on Georgian Bay
December 2004

Above: Gassing up, 40,000 gallons of Bunker C fuel oil
Below: Conveyorman's post, unloading Type II
Opposite: Christmas Eve, arriving at South Chicago
December 2004

Above is my favorite frame among the thousands I snapped during the *Centennial* project–a frigid and solitary Christmas Eve arrival alongside *C.T.C. No. 1* at South Chicago. In my head the tableau is always scored by the haunting measures of "Rock Island, 1931" from the *Road to Perdition* soundtrack. The guy in the jacket was waiting to deliver groceries for Christmas dinner the next day.

Above: Christmas morning, business as usual
Opposite: Shifting aft alongside *C.T.C. No. 1*
December 2004

Above: Dawn in the Straits, eastbound, the day after Christmas
Opposite: Table setting in the officer's mess, Christmas dinner 2004
December 2004

Above: Heavy weather loading at Charlevoix
Opposite left: Turning the boat in Milwaukee Harbor
Opposite right: New Year's Eve, downbound on Lake Michigan
January 2005

Above: Lay-up load, 9200 net tons of Type I, Trip No. 67, -8° F
Opposite: Bow view in the loading slip at Charlevoix
January 31, 2005

Above: Capping the smokestack at winter lay-up
Opposite: Engine room trip board, final load of the season
February 2005

Above: Engine crosshead block and connecting rod
Opposite: Checking tolerances on the piston rings
February 2005

Centennial

St. Marys Challenger
2005

Above: View of the fantail, the boat's 99th season and sixth name
Opposite: Work gang painting *St. Marys Challenger* on the starboard bow
June 2005

Above: The *Challenger*'s new pilothouse nameboards arrive
Opposite left: Resetting the masthead light during an unusually late fit-out
Opposite right: Blowing flues, bringing the engine room back to life
June 2005

Above and opposite: First trip as *St. Marys Challenger*
June 14-16, 2005

Above: Arriving at the mouth of the Grand River
Opposite: Oiler sounding diesel tanks, portside aft
June 2005

Above: Clearing the Five Bridges complex on the Calumet River
Opposite: Forward-enders headed aft for weekly fire and boat drill
July 2005

Trying to will a steamboat, an aircraft and my camera into proximity in the sweet light of a decent weather day shaved ten years off my life. I can barely imagine attempting this sort of photograph in an age before cell phones and the availability of real-time satellite weather information. The tropical splendor of the water in this vicinity of the lake made it easier to ignore the fact that I had been up for 30 hours, had nothing in my stomach, and was pretty airsick as a result.

Above and opposite: Arriving at St. Marys Charlevoix, Michigan cement plant
July 2005

Below: Unloading Type II at Ferrysburg
Opposite: Captain George Herdina
August 2005

Above: Passing the Lakes' last battery of Hulett Unloaders, South Chicago
Opposite: Silo-top view of *Challenger*'s after deckhouses
August 2005

Above: Posting the lunch menu in the crew's mess
Opposite: Porter at work in the galley prep area
August 2005

Above: Backing out of Grand Haven entry, Labor Day weekend
Opposite: Crossing northern Lake Michigan, westbound
September 2005

Above: Deckhand making up the boat prior to arrival at Charlevoix
Opposite: Hosing down after departing the loading slip
September 2005

Above: Outbound on the Calumet, approaching the Torrance Avenue Bridge
Opposite: The boat's forward end, view from the cement silo at South Chicago
September 2005

Above: Clearing the Soo Line jackknife bridge on the Manitowoc River
Opposite: View from the bridge wing, departing the *C.T.C. No. 1*
September 2005

Above: Fall colors above 10th Street on the Manitowoc River
Opposite: Passing steamer *Alpena* in Milwaukee's inner harbor
October 2005

Above: Passing the *Huron Lightship*, downbound on the upper St. Clair River
Opposite: Fall color at Fort Gratiot Light, downbound for Lafarge's Detroit silo
October 2005

**Above and opposite: A rare charter load from the Lafarge Corporation's Alpena, Michigan plant
October 2005**

Above: Pulling cable, departing the Sterling Fuel pier at Windsor, Ontario
Opposite: Twilight from the Manitowoc, Wisconsin, elevator, inbound
October 2005

Above: A winter's nap for local buoys, Charlevoix
Opposite: Deckands stowing boiler feedwater salt
November 2005

November 2005 was one of the nastiest weather months in recent history. Six major storm-force systems howled over the Lakes, idling vessels for days at a crack and costing the U.S. flagged fleet some 5000 hours of lost time. This storm on Thanksgiving Day came on like a switch—about five minutes after it was predicted to hit, the windows in the pilothouse began to moan. We stayed tucked safely in the loading slip with our turkey dinners and clocked a 55-knot gust *behind* the giant cement silos. Waves on Lake Michigan went to 25 feet—a fully developed sea state.

Above: Hurricane conditions at the Charlevoix breakwall, Thanksgiving Day
Opposite: The Great Lakes' fifth storm-force weather system of November 2005

Steak Night is a tradition as old as steamboating itself. Every Saturday night *Challenger*'s galley staff wrangles up a herd of sirloins big as manhole covers. Unfortunately, all the baked potatoes and sauteed mushrooms in the world can't completely assuage the grief of not being able to wash the meal down with a big 16 oz. beer. It seems the U.S. Coast Guard frowns on that sort of thing... ah well.

Below: Thanksgiving dinner, downbound on Lake Michigan
Opposite: Steward Sam Al-Samawi, Saturday Steak Night
November 2005

Above: The boat's reciprocating steam anchor windlass engine
Opposite: Departing South Chicago, outbound at 95th Street
December 2005

Above: Heavy weather on northern Lake Michigan
Opposite: Prepping for the dinner hour, South Chicago
December 2005

Above: Tug *Massachusetts* breaking out the slip
Opposite: Winter solstice on Lake Calumet, arriving with Type I
December 2005

Below: Inbound for winter lay-up at South Chicago
December 2005

A steamboat is never more alive than in subfreezing weather. There was no fast ice on the Calumet, and an early cold snap made its surface writhe with sea smoke. The transformative magic of light and steam are front and center here, bewitching a steely winter landscape that most morning visits would find grim at best.

Below: Oiler cleaning the boiler flues on New Year's Day
Opposite: Sledging open engine cylinder heads for winter work
January 2006

By early January, winter ice turns rivers and harbors to stone, and it becomes too difficult for the Lakes' shipping industry to function economically. The traditional winter lay-up contributes to a laker's renown life expectancy, as it permits her owners a unique two or three month window to tend to scheduled maintenance projects.

Above and opposite: Pulling pistons one and three for repair and inspection
January 2006

Centennial

Centennial Season

2006

Above: Lighting off the starboard boiler at spring fit-out
Opposite: Celebratory billboard lettering on the pilothouse
April 2006

200

Below: First load of the centennial season
Opposite: Trimming at Charlevoix, Trip No. 1
April 9, 2006

"So it is that the season passes... spring through summer through fall, and the ships pass too, upbound and downbound, all day and through the night, taking dawns in their stride. Pushing on past sunsets..."

—From the Lake Carriers' Association's
1960 short film *The Long Ships Passing*

Above left: Steaming off spring ice in the loading slip
Above right: DEU inspecting the unloading conveyor
Opposite: Distant traffic, upbound on Lake Michigan

Above: Deckhands bundled up for a cold shift under the loading spouts
Opposite: Oiler's rounds on the engine room Fidley Deck
April 2006

Above: Arriving at Charlevoix on the 100th anniversary of *Challenger*'s maiden voyage
Opposite: Loading 9200 tons of Type I for the boat's centennial voyage
0450 Friday, April 28, 2006

Challenger's centennial anniversary dawned clear as a bell, with the boat arriving in the sweetest light of the day. The hull was transformed into molten gold for ten solid minutes that morning, a phenomenon I hadn't witnessed in three dozen visits to the loading slip. Evidently, the fickle steamboat gods were pleased.

Below: A century in steam, and an unprecedented event in the history of Lakes' shipping
Opposite: 0930, Trip No. 8 of the 2006 season, *Challenger* trimmed and ready to eclipse the magic 100-year milestone
April 28, 2006

THIS SHIP TO SAIL AT 0900 ON 4-28

Below: An historic split load of Type I cement for the Ferrysburg and Milwaukee terminals
Opposite: Logging out the centennial voyage
April 29-30, 2006

Above: Returning home to the Charlevoix plant, Monday, May 1, 2006

Opposite: Bagging cement samples on a raw morning, another routine cargo for the most extraordinary of ordinary ships

Above: Moonrise at the Manitowoc North Breakwall Light
Opposite: Afterglow, downbound off Sheboygan, Wisconsin
May 2006

Above: Morning coffee break in the engine room
Opposite: Conveyorman Ron Bujokovsky boards the *St. Marys Express* in the conveyor tunnel
May 2006

Some lake carriers have golf carts to speed crewmen fore and aft along their block-long conveyor tunnels. True to form, *Challenger* boasts this ancient cement-baked Schwinn.

217

Below: Maneuvering, inbound on the Kinnickinnic River, fresh morning
Opposite: Grand Haven Outer Pierhead Light from the chartroom window
June 2006

Above: Taking a load off, crew rec, aft
Opposite: Trim time in the engine room
July 2006

Often, lake sailors don't bother cabbing it up the street to get their ears lowered. This struck me as a timeless scene, as I have meandered through at least 200 scrapbooks of shipboard pictures in museum archives over the years, and without fail every one of them includes the obligatory snaps of crewmen giving each other a trim.

Above: Outbound in light fog on the Manitowoc River, Trip No. 36
Opposite: View of the observation lounge from the starboard patio
July 2006

Above left: Porter baking fresh cinnamon rolls in the galley
Above right: Oiler counting drips on the Manzel Lubricator
Opposite: Clearing the Daniel Hoan Memorial Bridge at Milwaukee
July 2006

Challenger has been calling at the cement terminal on the Kinnickinnic River since her conversion in 1967. My grandfather and I used to fish along the entry channel to Milwaukee's inner harbor, and I have deep memories of being boosted up to try and get her skipper to blow a salute. Milwaukee was also the spot where we seemed to get stuck whenever the weather went to hell out on the lake. Over the course of five seasons I managed to spend at least a week marooned on the boat in my own backyard.

Above: On the winches at Charlevoix, loading Type I for South Chicago
Opposite: Mates John McNabb and Bill Kischel spotting booms prior to loading
September 2006

Above: Short hops, Milwaukee/Charlevoix/Chicago
Opposite: Rounding Sand Mine Bend on the Grand River
September 2006

Below and opposite:
Long haul poker in the crew rec, aft
September 2006

Above: View of the unload from the former Manitowoc Shipbuilding Company yard
Opposite: Fueling via tanker *William L. Warner* at South Chicago
October 2006

I gambled on this dirty weather day... and lost. By noon the boat was pinned down in Grand Traverse Bay with a MAFOR calling for winds in excess of 40 knots for the next six days. I had major project deadlines on the bubble at work, and *had* to get off the boat. After 24 hours of frantic phone calling I was put off in a fisherman's Zodiac, and proceeded to endure a frigid 14-mile slog down to Traverse City. By the time we tied up, my camera bags had to be chipped out of an inch-thick carapace of ice.

Opposite: Relief Captain Kenny Lichtle in the window, departing the slip at Charlevoix
Below: Foul weather on Lake Michigan, downbound with a lay-up load for South Chicago
December 2006

Above and opposite: Arriving at the Chicago Seaway Terminal for winter lay-up
December 18, 2006

Hull No.17/*St. Marys Challenger* WDB9135

Current ownership: St. Marys Cement (US) Incorporated, Detroit, Michigan (division of St. Marys Cement Incorporated). Vessels operated & managed by HMC Ship Management Limited, affiliate of Hannah Marine Corporation, Lemont, Illinois

Official number: 202859

Built: Great Lakes Engineering Works at Ecorse, Michigan, launched February 17, 1906 as the *William P. Snyder*. Departed on maiden voyage April 28, 1906

Length: 552' 1"

Beam: 56'

Depth: 31'

Light ship weight: 5,400 tons

Total displacement, mid-summer: 17,720 tons

Cargo capacity at mid-summer draft: 66,240 barrels of cement

Maximum unloading rate: 8,000 barrels per hour

Engine/Horsepower–original: 1,665 i.h.p. triple-expansion steam engine, built by the Great Lakes Engineering Works, at Ecorse, Michigan

Engine/Horsepower–current: 3,500 i.h.p. 4-cylinder simple-type Skinner Marine Unaflow steam engine, built by the Skinner Engine Company, at Erie, Pennsylvania

Vessel Lineage:

- 1906-1926: ***William P. Snyder*** Shenango Furnace Company, Cleveland, Ohio
- 1926-1929: ***Elton Hoyt II*** Stewart Furnace Company, Cleveland, Ohio
- 1929-1930: ***Elton Hoyt II*** Youngstown Steamship Company, Cleveland, Ohio
- 1930-1952: ***Elton Hoyt II*** Interlake Steamship Company, Cleveland, Ohio
- 1952-1966: ***Alex D. Chisholm*** Interlake Steamship Company Cleveland, Ohio
- 1966-1999: ***Medusa Challenger*** Cement Transit Company, Division Medusa Portland Cement, Detroit, Michigan
- 1999-2005: ***Southdown Challenger*** Southdown Incorporated, then Wilmington Trust, Delaware, operated by HMC Ship Management Limited, Lemont, Illinois
- 2005-2007: ***St Marys Challenger*** St. Marys Incorporated, Detroit, Michigan; operated by HMC Ship Management Limited, Lemont, Illinois

Archival Photographic and Illustration Sources

All photos by the author unless otherwise credited. Every effort has been made to attribute material reproduced in this book correctly. If any errors have occurred we will be happy to correct them in future editions.

Front dust jacket flap: Postcard of the *William P. Snyder* departing the Canadian Sault–Paul LaMarre III Collection
Endpaper front: Illustration of an ore boat under a battery of Hullet unloaders from *This is Shenango*, 1954–Author's Collection
Endpaper back: The *William P. Snyder* on the drydock at the Great Lakes Engineering Works, 1906–Library of Congress/Detroit Publishing Company Collection
Copyright page: Navigation chart of Lake Michigan, 1906–From the Image Archives of the Historical Map & Chart Collection/Office of Coast Survey/National Ocean Service/NOAA.
Page 4: The Shenango fleet rafted in Cleveland Harbor April 23, 1917, photographer unknown–Author's Collection
Page 6: *Medusa Challenger* jacket patch–Groh Collection
Page 8: Pen and ink drawing of the *Medusa Challenger* clearing the Chicago Skyway Bridge–Artist unknown–Larry Geiger Collection
Pages 11 & 14: Charcoal sketches from *This is Shenango*, 1954–Author's Collection
Pages 18 & 23: Pencil sketches of the *St. Marys Challenger* by Bill Moss
Pages 24-25: The *William P. Snyder* at the Great Lakes Engineering Works, 1906–Library of Congress/Detroit Publishing Company Collection
Pages 26-27: Excerpts from *The Marine Review* of 1906–Great Lakes Historical Society Collection
Pages 28-29: Interior and profile photographs of the *Snyder*, photographer unknown–Author's Collection. Photograph of Captain Henry Peterson from *The Marine Review* of 1906–Great Lakes Historical Society Collection. *Marine News* item reprinted with permission of the Detroit Free Press
Pages 30-31: The *William P. Snyder's* observation lounge, photographer unknown–Author's Collection. Blueprints of the *Snyder's* Texas Deck–John Belliveau Collection
Page 32-33: The *William P. Snyder's* triple-expansion engine, photographer unknown–Author's Collection. The Shenango fleet upbound at the Soo Locks, photograph by A.E. Young–Paul LaMarre III Collection
Pages 34-35: The *William P. Snyder* upbound at the Poe Lock at Sault Ste. Marie, Michigan, photographer unknown–Wisconsin Marine Historical Society/MPL Collection
Pages 36-37: Article reprinted with permission of the Milwaukee Journal Sentinel. Captain J.J. Slade, photographer unknown–Author's Collection
Pages 38-39: *Elton Hoyt II* repowering project photographs, photographer unknown–Door County Maritime Museum Collection. Skinner Engine Company advertisement–Art Chavez Collection
Pages 40-41: Photo of the *Alex D. Chisholm* passing Detroit, photographer unknown–Wisconsin Maritime Museum Collection. Aerial view of the *Chisholm* crossing Lake St. Clair, photographer unknown–Fr. Edward Dowling Collection, University of Detroit–Mercy. Fit-out as *Elton Hoyt II*, photographer unknown–Fr. Edward Dowling Collection, University of Detroit–Mercy. *Alex D. Chisholm* downbound on the Detroit River, photograph by Scott B. Worden Jr.–Worden Family Collection
Pages 42-43: The *Medusa Challenger* at the Manitowoc Shipbuilding Company, photograph by Demisen Studio–Author's Collection. The *Medusa Challenger* during re-christening ceremony at Cleveland, Ohio, photographer unknown–Author's Collection. Re-christening admission card–Author's Collection. Re-christening news item © 1967 Cleveland Plain Dealer. All rights reserved. Reprinted with permission.
Pages 44-45: *Medusa Challenger* inbound on the Chicago River, 1970–photographs by James Bartke. News item © 1969 Chicago Tribune Company. All rights reserved. Used with permission.
Pages 46-47: Steward and port chaplain, photographer unknown–Larry Geiger Collection. Menu from Thanksgiving, 1976–Larry Geiger Collection. Postcard of *Medusa Challenger* in bicentennial livery–Author's Collection. News item © 1977 Chicago Tribune Company. All rights reserved. Used with permission. Snapshots of lake rescues by Larry Geiger–Larry Geiger Collection
Pages 48-49: *Welcome Aboard the Medusa Challenger* brochure–Author's Collection. *Medusa Challenger* jacket patch and playing cards–Groh Collection Illustration of vessel lineage by John Belliveau–www.boatnerd.com/digitalshipyard
Pages 53-143-197: Pencil sketches of the *St. Marys Challenger* by Bill Moss
Endpaper front: Postcard detail of the *William P. Snyder* under the Huletts at Cleveland, Ohio–Paul LaMarre III Collection